Technical Assessment of the Man-In-Simulant Test (MIST) Program

Report 1

ASSESSMENT OF THE U.S. ARMY CHEMICAL AND BIOLOGICAL DEFENSE COMMAND

Standing Committee on Program and Technical Review of the
U.S. Army Chemical and Biological Defense Command
Board on Army Science and Technology
Commission on Engineering and Technical Systems
National Research Council

NATIONAL ACADEMY PRESS
Washington, D.C. 1997

NOTICE: The project that is the subject of this report was approved by the Governing Board of the National Research Council, whose members are drawn from the councils of the National Academy of Sciences, the National Academy of Engineering, and the Institute of Medicine. The members of the committee responsible for the report were chosen for their special competencies and with regard for appropriate balance.

This report has been reviewed by a group other than the authors according to procedures approved by a Report Review Committee consisting of members of the National Academy of Sciences, the National Academy of Engineering, and the Institute of Medicine.

The National Academy of Sciences is a private, nonprofit, self-perpetuating society of distinguished scholars engaged in scientific and engineering research, dedicated to the furtherance of science and technology and to their use for the general welfare. Upon the authority of the charter granted to it by the Congress in 1863, the Academy has a mandate that requires it to advise the federal government on scientific and technical matters. Dr. Bruce Alberts is president of the National Academy of Sciences.

The National Academy of Engineering was established in 1964, under the charter of the National Academy of Sciences, as a parallel organization of outstanding engineers. It is autonomous in its administration and in the selection of its members, sharing with the National Academy of Sciences the responsibility for advising the federal government. The National Academy of Engineering also sponsors engineering programs aimed at meeting national needs, encourages education and research, and recognizes the superior achievements of engineers. Dr. William A. Wulf is interim president of the National Academy of Engineering.

The Institute of Medicine was established in 1970 by the National Academy of Sciences to secure the services of eminent members of appropriate professions in the examination of policy matters pertaining to the health of the public. The Institute acts under the responsibility given to the National Academy of Sciences by its congressional charter to be an adviser to the federal government and, upon its own initiative, to identify issues of medical care, research, and education. Dr. Kenneth I. Shine is president of the Institute of Medicine.

The National Research Council was organized by the National Academy of Sciences in 1916 to associate the broad community of science and technology with the Academy's purposes of furthering knowledge and advising the federal government. Functioning in accordance with general policies determined by the Academy, the Council has become the principal operating agency of both the National Academy of Sciences and the National Academy of Engineering in providing services to the government, the public, and the scientific and engineering communities. The council is administered jointly by both Academies and the Institute of Medicine. Dr. Bruce M. Alberts and Dr. William A. Wulf are chairman and vice chairman, respectively, of the National Research Council.

This is a report of work supported by Contract DAAM01-96-K-0002 between the U.S. Army Chemical and Biological Defense Command, and the National Academy of Sciences.

Any opinions, findings, conclusions, or recommendations expressed in this publication are those of the author(s) and do not necessarily reflect the view of the organizations or agencies that provided support for the project.

International Standard Book Number 0-309-05940-2

Limited copies are available from:
Board on Army Science and Technology
National Research Council
2101 Constitution Avenue, N.W.
Washington, DC 20418

*Additional copies are available for sale from:*National Academy Press Box 285 2101 Constitution Ave., N.W. Washington, DC 20055 800-624-6242 202-334-3313 (in the Washington Metropolitan Area)

Printed in the United States of America.

Standing Committee on Program and Technical Review of the U.S. Army Chemical and Biological Defense Command

FRANCIS G. DWYER *(chair)*, Mobil Research and Development Corporation (retired), West Chester, Pennsylvania
JEROME S. SCHULTZ *(vice chair)*, University of Pittsburgh, Pittsburgh, Pennsylvania
KLAUS BIEMANN, Massachusetts Institute of Technology, Cambridge
HAROLD S. BLACKMAN, Lockheed Martin Idaho Technologies Company, Idaho Falls, Idaho
BARBARA G. CALLAHAN, Fluor Daniel GTI, Norwood, Massachusetts
JAMES M. CARUTHERS, Purdue University, West Lafayette, Indiana
CLEMENT E. FURLONG, University of Washington, Seattle
JEAN H. FUTRELL, University of Delaware, Newark
BERNARD D. GOLDSTEIN, Robert Wood Johnson Medical School, Piscataway, New Jersey
MICHAEL T. KLEIN, University of Delaware, Newark
SANFORD S. LEFFINGWELL, HLM Consultants, Dacula, Georgia
MAJOR GENERAL WILLIAM L. MOORE, JR., U.S. Army (retired), Nashville, Tennessee
DEREK L. RANSLEY, Chevron Research and Technology Company (retired), Richmond, California
LUDWIG REBENFELD, TRI Princeton, Princeton, New Jersey
WILLIAM REIFENRATH, Reifenrath Consulting & Research, Richmond, California
K. ANNE STREET, DynCorp, Reston, Virginia
J. THROCK WATSON, Michigan State University, East Lansing

Board on Army Science and Technology Liaison

CHRISTOPHER C. GREEN, General Motors Corporation, Warren, Michigan

Army Liaisons

MAJOR GENERAL GEORGE E. FRIEL, U.S. Army Chemical and Biological Defense Command (CBDCOM), Aberdeen Proving Ground, Maryland
JOSEPH VERVIER, U.S. Army Edgewood Research, Development and Engineering Center, Aberdeen Proving Ground, Maryland
JAMES BAKER, U.S. Army Edgewood Research, Development and Engineering Center, Aberdeen Proving Ground, Maryland

Staff

BRUCE A. BRAUN, Director, Board on Army Science and Technology
ERIC SHIMOMURA, Senior Program Officer (to March 1997)
GEORGE DAVATELIS, Senior Program Officer
JACQUELINE A. CAMPBELL-JOHNSON, Senior Program Assistant

BOARD ON ARMY SCIENCE AND TECHNOLOGY

Preface

This report is the first of a two-phase response to a request from the technical director of the U.S. Army Edgewood Research, Development and Engineering Center (RDEC) that the National Research Council's (NRC) U.S. Army Chemical and Biological Defense Command Standing Committee (CSC) conduct technical assessments and program reviews within the command. Specifically, the CSC was asked to conduct a technical assessment of the man-in-simulant test (MIST) program and a program review of the mass spectrometry and bioremediation programs. These programs represent a continuum of technologies designed to protect, detect, and dispose of chemical and biological weapons that soldiers may face in future combat. This report focuses on the technical assessment of the MIST program.

Members of the CSC have a wide range of expertise in chemical engineering, chemistry and biochemistry, toxicology and risk assessment, simulation and modeling, bioremediation of chemical warfare agents, physical chemistry and mass spectrometry, medicine, chemical modeling, epidemiology and industrial hazards, and military science. Members of the committee whose expertise was relevant to reviewing the MIST program were chosen to serve on the review panel. The panel met three times between October 1996 and April 1997 and heard testimony from several Army research and development experts, including representatives from the Edgewood RDEC, the U.S. Army Center for Health Promotion and Preventive Medicine, the Natick RDEC, and Dugway Proving Ground, in Utah, where the tests are conducted.

In this report, the committee documents the methodology used by the Army to test protective suit ensembles and analyze data. The committee carefully considered the best way to present its findings and organize the report, given the critical nature of the MIST program and its ramifications for Army personnel. The problem is complicated by the fact that the Edgewood RDEC is faced with operating in an environment of constrained defense budgets and reductions in military and civilian personnel. The Edgewood RDEC's workforce has been reduced by 20 percent since 1990, and the U.S. Army Material Command projects another 15 percent reduction by 2000. Funding that had been earmarked for defense research and development is also being transferred to military operations. These reductions in personnel and funding will require that priorities be precisely determined and that data be generated efficiently. To that end, the technical director of the Edgewood RDEC requested that the NRC provide expert, independent technical advice and counsel on selected aspects of the nuclear, biological, and chemical research, development, and acquisition program. The chair and the committee wish to express their gratitude for the staff assistance and support provided by the NRC. We are indebted to Bruce Braun, director, Board on Army Science and Technology; George Davatelis, study director; Jacqueline Campbell-Johnson, senior project assistant; Margo Francesco, staff associate; Alvera Gircys, financial associate; and William Holm, consultant. The work of the committee would not have been possible without these dedicated individuals. The committee also appreciates the comments and written submissions of the various groups who provided testimony and written material; Virginia Gildengorin, for reviewing the data analysis procedures; and the group of outside experts who graciously donated their time to review this report.

Francis G. Dwyer, *chair*

Standing Committee on Program and Technical Review of the U.S. Army Chemical and Biological Defense Command

Contents

Tables and Figures

TABLES

FIGURES

Acronyms

ANOVA	analysis of variance
BRHA	body region hazard analysis
CB	chemical and/or biological agent
CBDCOM	U.S. Army Chemical and Biological Defense Command
CPE	chemical protective ensemble
CSC	CBDCOM Standing Committee
Ct	concentration x time
CWA	chemical warfare agent
GA	nerve agent (chemical warfare agent)
GB	nerve agent (chemical warfare agent)
H or HD	mustard, blister agent (chemical warfare agent)
HDPE	high-density polyethylene
MeS	methyl salicylate
MIRANS	miniature infrared analyzers
MIST	man-in-simulant test
NBC	nuclear, biological, and chemical
NRC	National Research Council
PF	protection factor
RDEC	Research, Development and Engineering Center
VX	nerve agent (chemical warfare agent)
WBEE	whole body effective exposure

Executive Summary

In 1993 the U.S. Army established the Chemical and Biological Defense Command (CBDCOM) to conduct research, develop and procure support systems, and design equipment to protect U.S. military personnel from the increasing threat by foreign entities and terrorist organizations to use chemical and biological weapons. CBDCOM is the latest in a long history of military organizations designated for chemical and biological defense research. Because of the critical nature of its mission, the CBDCOM requested that the National Research Council (NRC) establish an oversight committee of nationally recognized experts to provide ongoing, impartial, independent advice and assessments.

The NRC, responding through the Board on Army Science and Technology of the Commission on Engineering and Technical Systems, created a standing committee called the Program and Technical Review of the U.S. Army Chemical and Biological Defense Command, better known as the CBDCOM Standing Committee (CSC). This committee was assembled to provide expertise in the areas of science and technology pertinent to the concerns of the CBDCOM commander and executive director and the technical director of the Edgewood Research, Development and Engineering Center (RDEC), which historically has been an important organization in the Army and Department of Defense for chemical and biological research.

The U.S Army has not established specific requirements for the chemical protective qualities of its ensembles (chemical protective ensembles, or CPEs). This is because test results (protection factors) have never been correlated with biological endpoints. Instead, new CPEs have been evaluated in comparison to the CPE currently in the field (e.g., the battle dress overgarment, BDO). The goal of the Army's program has been to increase chemical protection factors while decreasing undesirable properties (weight and heat stress), although there is no clear understanding of how much chemical protection would be enough to maintain battlefield effectiveness. The man-in-simulant test program (MIST) is responsible for specifying protection factors, but it does not, by itself, link them to biological effects and has not answered the CPE developer's question of how much protection is enough.

The cornerstone of chemical and biological defense strategy is protection (i.e., insulating personnel from chemical and biological agents using individual clothing ensembles and respirators, as well as collective filtration systems and shelters). The CSC was asked by the CBDCOM to undertake a technology assessment of the Army's MIST program—which is designed to test protective suit ensembles in simulated chemical attacks. Specifically, the CSC was asked to:

1. review the test methodology for the man-in-simulant test program[1]
2. review the use of biological markers (e.g., cholinesterase inhibition) to predict the signs and symptoms associated with exposure to nerve (VX) and vesicant (HD) agents
3. review the test methodology for employing passive and active vapor and aerosol samplers during simulant tests at Dugway Proving Ground, Utah, and assess the plan for data collection and analysis
4. determine whether the current chemical simulant, methyl salicylate, or an alternative simulant should be used in the MIST program

To accomplish this task the CSC established a panel of experts from members of the committee to undertake the MIST review. The panel has addressed each item on the list and has summarized the conclusions and recommendations below. The background information and rationale behind these findings are detailed in the full report.

[1] The original statement of task for Task 1 included "and the rationale for using methyl salicylate as a chemical agent simulant in this test program." The committee felt that this aspect of the review was reiterated in Task 4 and has addressed the question there.

TASK 1. Review the test methodology for the man-in-simulant test program.

Conclusion 1. The MIST is a well-designed test protocol for evaluating chemical protective ensembles. However, the committee found that the test methodology was not based on preliminary testing that would eliminate ensembles with gross defects and allow more replications of tests be done on fewer candidate protective ensembles, thereby increasing the statistical power of the results.

Recommendation 1. The Army should screen ensembles prior to a full-blown MIST by video imaging the skin of test subjects after exposure to a fluorescent tracer or other physical tests. Screening should also include variations in ambient conditions (temperature, humidity, wind, and, perhaps, rain), activities (kneeling, sitting, and crawling), and sweat-soaked and dry test challenges.

TASK 2. Review the use of biological markers (e.g., cholinesterase inhibition) to predict the signs and symptoms associated with exposure to nerve (VX) and vesicant (HD) agents.

Conclusion 2. Body region hazard analysis (BRHA) is an innovative approach that takes into account regional variations in skin sensitivity to chemical agents. Although the basic approach is sound, the committee has the following reservations:

- A direct relationship has not been established between cholinesterase depression and the percutaneous absorption of agent.
- The relationship between liquid and vapor absorption has not been determined.
- BRHA was based on the local absorption of VX and may not accurately predict the absorption of HD.
- BRHA does not account for functional impairments from mustard-induced lesions in various body regions.
- BRHA does not account for individual differences in sensitivity to chemical agents.

A direct determinant of the toxicity of a chemical agent is the permeability of the skin by that agent at a given anatomic site.

Therefore, the committee concluded that rather than basing the BRHA on highly variable indirect measures (cholinesterase depression) and assumptions, a protocol should be designed to quantify the *in vitro* agent permeability of excised human skin samples from different body regions. These techniques are well established and well accepted and could also be used to compare simulant uptake by human skin and passive samplers. Large differences may indicate a need to redesign the samplers. The vapor uptake of agent and simulant could also be determined for human skin and passive samplers. Large differences in the behavior of agent and simulant may warrant the selection of a different simulant or adjustments in the methods used to calculate protection factors.

Recommendation 2a. The Army should measure regional variations in skin penetration for HD, VX, and simulant vapors using excised human skin harvested from various anatomic sites.

Recommendation 2b. As a supplemental validation of the systematic BRHA, a biomonitoring protocol should be developed for the MIST, analogous to the protocol used to monitor pesticide exposures to agricultural workers. If the appropriate simulant is used, the calibrations obtained from *in vitro* studies could be used to relate suit performance to physiological effects based on the absorbed dose.

TASK 3. Review the test methodology for employing passive and active vapor and aerosol samplers during simulant tests and assess the data collection and analysis plan.

Conclusion 3. Passive samplers are appropriate means for testing for the presence of vapor. The protocol, however, may not be valid for aerosols because the disposition of chemical agents in aerosol and vapor forms can be quite different. From the information recorded in the documents given to the committee for review, the committee could not confirm the uniformity of simulant concentration within the test chamber. Variations in concentration outside the protective ensemble could lead to errors in assessing the protective qualities of the suit.

Although passive samplers are generally regarded as less accurate than active samplers in bench trials, the differences in the results are small. The precision and accuracy of the Natick sampler is adequate for the intended purpose. The small size of the Natick sampler enables testing under the suit without incurring a number of disadvantages (outlined in Chapter 4) that would be incurred with active sampler pumps either inside or outside the suit.

A residual disadvantage of passive samplers may be a lack of sensitivity to brief variations in concentration, which would be of interest only for identifying the body positions or activities associated with leakage. Conventional active samplers would have the same disadvantage, but external samplers connected to a near-real-time monitor could provide this information.

Recommendation 3. Agent uniformity in all parts of the test chamber throughout the duration of the tests should be documented. In addition, concentrations inside the suit could be monitored with either active or passive samplers, despite their logistical problems. Comparing simulant levels in the passive sampler with samples recovered from the stratum corneum of test subjects (the outermost layer of the skin, which can be removed by repeated applications of adhesive tape) would provide insights into sampler performance.

TASK 4. Determine whether the current chemical simulant, methyl salicylate, or an alternative simulant should be used in the MIST program.

Conclusion 4. Methyl salicylate is an appropriate simulant for the transport of chemical agent into protective ensembles. However, biological interpretations of the MIST/BRHA using methyl salicylate are not warranted.

Recommendation 4. Additional studies should be undertaken to establish absorption and transport properties of the simulant relative to the properties of the agents. *In vitro* studies using excised skin and mannequin studies (capable of simulating a bellows effect) can be used to accomplish this objective. With the appropriate consent and the oversight of a human use committee, excised human skin can be used for research. Samples can be obtained from cadavers or from surgical samples (e.g., abdominal skin, facial skin, etc.) Large differences in distributions may warrant that an alternative simulant be used.

GENERAL CONCLUSIONS AND RECOMMENDATIONS

General Conclusion 1. The first step in chemical and biological defense strategy is early detection and warning to provide situational awareness and permit steps to be taken to avoid the exposure of personnel and equipment. The complement to detection is protection. Chemical protective ensembles, as well as collective filtration systems and shelters, are used to insulate personnel from chemical and biological agents. Modeling chemical protective ensembles is a daunting task, and the Army's efforts to develop the MIST/BRHA should be commended. Modeling and simulation technologies are invaluable tools for training for operations in a chemical and biological warfare environment. They provide material and equipment design parameters and enable field commanders to integrate and interpret real-time data. However, deriving physiological endpoints from the MIST/BRHA is a complicated process that will require cooperation among the Army's scientists, as well as significant input from academia and industry.

General Recommendation 1. The development of new test methodologies should be done separately from routine ensemble testing. Once the criteria for suit performance have been established, decision points should be entered in a flow chart to reveal where additional work is needed. As of this writing, the Army has not adopted a clear approach to establishing physiologic endpoints from protective ensemble testing. However, this is an achievable goal that should be pursued to protect soldiers

General Conclusion 2. The Army should ensure better cooperation among various disciplines (i.e., chemistry, toxicology, engineering, human factors, etc.). For example, scientists in CBDCOM's toxicology division have not participated in any significant way in the development of ensemble test methods.

General Recommendation 2. More integration between the various groups and technical disciplines will be essential for the development of future testing methodologies. All relevant parties should participate in the planning phase with the objective of reaching a consensus on research objectives, design procedures, analysis, and documentation.

1

Introduction

BACKGROUND

The purpose of chemical and biological (CB) defense research is to develop equipment that will protect U.S. military forces, sustain combat operations, and maintain system effectiveness in a CB agent-contaminated environment. The cornerstone of a CB defense strategy is early detection and warning to provide situational awareness and time to take steps to avoid the exposure of personnel and equipment. The complement to detection is protection (i.e., to insulate personnel from CB agents using individual clothing ensembles and respirators, as well as collective filtration systems and shelters). Modeling and simulation technologies are used to assess conditions, train personnel, develop material for operating in a CB warfare environment, provide equipment design parameters, and enable field commanders to integrate and interpret real-time data.

In 1993 the Army established the U.S. Army Chemical and Biological Defense Command (CBDCOM), which is responsible for nuclear, biological, and chemical (NBC) defense, technology, products, and services to support U.S. forces, ensure the safe storage of chemical material, oversee the remediation and restoration of areas after exposure, and support chemical treaties and demilitarization. The Edgewood Research, Development and Engineering Center (RDEC) supports CBDCOM by performing basic research and development for NBC defense programs for the Army.

In 1995, the CBDCOM requested that the National Research Council (NRC), through the Board on Army Science and Technology of the Commission on Engineering and Technical Systems, provide expert, impartial, independent advice. In response to this request, the NRC organized a standing committee called the Program and Technical

Review of the U.S. Army Chemical and Biological Defense Command, referred to here as the CBDCOM Standing Committee (CSC). This committee was assembled to provide expertise in areas of technology pertinent to CBDCOM's mission, which includes five primary areas:

- maintaining a chemical and biological defense technology base and procurement capability
- accurately relating the results of tests on chemical and biological defense equipment to battlefield performance
- responding to the Army, Congress, and the public about chemical and biological issues
- transferring defense technology
- interacting with the Army's battle laboratories and integrating its technology and advanced concepts

The CSC was asked to consider technology issues and systems to assist CBDCOM in defining a vision for the future. During its first year, the CSC was also asked to investigate potential studies that would address the concerns of the CBDCOM commander and executive director and the technical director of the Edgewood RDEC, which has historically been an important organization in the Army and U.S. Department of Defense for chemical and biological research.

After numerous visits and interviews with key personnel at Edgewood RDEC and CBDCOM and internal deliberations, the CSC focused on two major areas: (1) a technical assessment of the man-in-simulant test (MIST) program; and (2) a program assessment of the technical quality of the Edgewood RDEC's mass spectrometry and bioremediation programs. It was decided that the CSC would be split into two panels of relevant experts to address the two tasks. This report, the first of the two-phase response, presents an evaluation of test methodology and data assessment associated with the MIST program.

CHARGE TO THE COMMITTEE

The MIST was designed to test chemical protective ensembles for soldiers in the field. The MIST is part of a program that includes designing protective suits, developing test methodology, conducting tests, and conducting health hazard analyses. The CSC was asked to make a technical assessment of the MIST program. Specifically, the CSC was asked to:

1. review the test methodology for the MIST program[1]
2. review the use of biological markers (e.g., cholinesterase inhibition) to predict the signs and symptoms associated with exposure to nerve (VX) and vesicant (HD) agents
3. review the test methodology for employing passive and active vapor and aerosol samplers during simulant tests at Dugway Proving Ground, Utah, and assess the plan for data collection and analysis
4. determine whether the current chemical simulant, methyl salicylate, or an alternative simulant should be used in the MIST program

STUDY APPROACH

The CSC selected members of the standing committee to serve on a panel to review the MIST program. The panel was composed of experts in the fields of protective systems, toxicology, risk assessment, environmental and occupational health, simulation and modeling, textile science, human factors, organic chemistry, and chemical engineering. The panel collected data to assess the methodology used in the MIST for suitability, validity, and thoroughness. The panel received input from a variety of sources, including personnel from the Natick RDEC, where the suits were designed; the Edgewood RDEC, where the test methodology was developed; the U.S. Army Center for Health Promotion and Preventive Medicine, where the health hazard assessment was performed; and the West Desert Test Center at Dugway Proving Ground, where the tests were actually performed. The panel also heard perspectives from the Test and Evaluation Command at Aberdeen Proving Ground, Maryland; the NBC equipment manager from Quantico Marine Base in Quantico, Virginia; and program management perspectives from a representative of Fort Belvoir, Virginia. The panel reviewed the selection processes used to arrive at the current methodology for samplers, modeling, data collection, biomarkers, the extrapolation of data to human use, and all assumptions.

[1] The original statement of task for Task 1 included "and the rationale for using methyl salicylate as a chemical agent simulant in this test program." The committee felt that this aspect of the review was reiterated in Task 4 and has addressed the question there.

The panel collected the data, evaluated them for suitability, and came to conclusions about the model as applied to the testing of chemical protective suits.

This report summarizes the activities and the recommendations based on the CSC's review of the MIST program. Chapter 2 focuses on the test protocol; Chapter 3 reviews the simulant selection; Chapter 4 focuses on test methods and sampler selection; Chapter 5 outlines assumptions and limitations; and Chapter 6 presents the committee's conclusions and recommendations.

2

Test Protocol

UNDERLYING CONCEPTS

Introduction

Protecting an individual against CB agents requires a suit that not only protects against the external threat of CB agents in both liquid and vapor form, but also allows body moisture and heat transfer to the external environment from the microclimate between the body and the clothing. Thus, the materials in an effective suit must meet stringent performance demands and will be necessarily complex. The currently accepted CB protective suit, referred to as the battle dress overgarment, is constructed from a permeable, multilayered material that incorporates activated charcoal to absorb chemical agents. A permeable outer layer of fabric (a tri-blend of 58 percent cotton, 27 percent Kevlar aramid, and 15 percent nylon twill weave) is backed by a charcoal-loaded foam or nonwoven layer. The surface of the outer layer is coated with formulations containing fluoropolymers to minimize surface energy, thereby preventing wetting and wicking of aqueous and organic liquid chemical agents. Protection against vapor chemical agents is provided by the absorbing charcoal. The multilayered material is permeable to water vapor so that some body moisture (perspiration) can escape from the body microclimate; however, thermal stress to the individual is a significant limitation of this suit. To prevent liquid perspiration, which can poison the activated charcoal, from coming into contact with the charcoal-containing foam, a permeable, nonwettable inner fabric is inserted.

The committee recognizes that protection against chemicals is just one of several factors involved in the overall evaluation of a chemical protective ensemble; heat stress is another. Because the committee was charged with reviewing the MIST program only for how well the program evaluates the chemical protection of an ensemble, other factors, such as heat stress, are not addressed directly.

Materials Evaluation

Reliable testing and evaluation procedures are essential in the constant search for better protective suits. A battery of laboratory tests have been developed and are available to evaluate the performance of candidate materials. These tests include liquid and vapor chemical agent permeation tests through fabric samples mounted on diffusion cells. These so-called "cup tests" are based on measurements made under controlled laboratory conditions and are valid only for homogeneous, continuous materials. They are designed to make precise, accurate evaluations of both the protective aspects and the heat and moisture transfer characteristics of materials. These tests utilize swatches of material and are designed only for continuous, uninterrupted materials that are globally homogeneous and uniform. Other tests are used to evaluate the mechanical properties of materials, which must meet standard requirements for strength, tear resistance, abrasion resistance, bending flexibility, and similar characteristics.

Suit Technology

When a protective suit is constructed from a suitable material, the final product can no longer be considered homogeneous and continuous. In other words, its performance may no longer meet the required component performance criteria despite the fact that the material from which it was constructed meets all of the established performance standards. In contrast to a homogeneous and continuous material, which is planar or two dimensional, the final product is three dimensional, discontinuous, and otherwise structurally complex. The suit is fabricated from many panels that are stitched, bonded, or otherwise held together, which creates discontinuities. In addition, the suit must be integrated with other protective gear, such as a hood, a mask, gloves, and boots, which create additional discontinuities in the overall ensemble.

Furthermore, a protective system is required to function under dynamic conditions. The individual must be able to move and perform a variety of tasks. Physical motion creates local deformations of the suit material that can, under certain conditions, reach relatively high strains. These deformations alter the pore structure (shape and dimensions) of the material and modify both the barrier and the permeation properties. The physical motion may especially affect the discontinuities in the suit created by the stitching and bonding of component panels and by interfaces with other protective gear. In addition, CB protective suits may also be subject to wear and damage during use, potentially under extreme battlefield conditions. Perforations, punctures, and tears in the suit material will create further discontinuities.

Evaluation of Suit Systems

For the reasons outlined above, evaluating the performance of a CB protective ensemble under realistic, dynamic conditions is far more complex than evaluating component materials and does not lend itself readily to standard, controlled laboratory measurements. In recognition of this fact, the MIST program has been developed to evaluate individual CB protective ensembles under realistic, dynamic conditions. The essential elements of the MIST involve placing individuals wearing candidate suits in an enclosed environment (referred to as a defensive test chamber) containing a nonhazardous chemical compound (methyl salicylate vapor) that is intended to simulate a chemical agent. Dynamic conditions are created by individuals performing various physical tasks according to a specified time schedule. The test individuals are outfitted with sensing patches on their skin, referred to as passive sampling detectors, that absorb the chemical compound when and if it penetrates the protective suit system. The sensing patches are positioned at various places on the body and are analyzed at the conclusion of the test procedure. The raw data from the patch analyses are interpreted in terms of a model referred to as body region hazard analysis (BRHA). This model is based on the varying absorptive capacities of the skin in different regions of the body and incorporates significant assumptions to determine the overall protection factor.

Together, BRHA and the MIST attempt to provide a quantitative measure of the relative effectiveness of a protective suit system under realistic dynamic conditions. The MIST/BRHA must be viewed as a procedure for evaluating the performance of a complex system, intended not only to characterize overall performance, but also to identify the weakest elements in the system (probably the discontinuities discussed above).

METHODOLOGY

Methods for Evaluating Protective Clothing

Protective materials traditionally have been evaluated using physical approaches in which the amount of chemical that passes through the material is measured directly. The internally absorbed dose is calculated based on estimates of skin deposition and percutaneous transport. A relatively new approach to estimating the ability of clothing to protect humans from hazardous chemicals is biological monitoring. Biological monitoring usually involves a urinalysis to determine the amount of chemical or metabolite in the urine following an exposure. If the proportion of the absorbed dose excreted in urine is known through control studies, the urine level can be used to estimate the total absorbed dose. Biological monitoring is now being widely used to assess worker exposure to pesticides and other hazardous chemicals found in the workplace and has become a benchmark for standardizing other procedures for assessing worker exposure (Wang et al., 1989).

For obvious reasons, biological monitoring cannot be used directly in human studies to evaluate protective materials against CB agents. It can be used, however, with nontoxic agent simulants, as long as the comparative penetration rates (ensemble material and discontinuities, as well as skin) and other biokinetic factors of CB agents and simulants are known.

In addition to biological monitoring, several other techniques are available to evaluate the effectiveness of protective suits in the MIST scenario. These include air sampling, passive dosimetry, and fluorescence imaging. Briefly, air monitoring involves sampling the air between the skin and the protective clothing. Passive dosimetry involves placing patches between the skin and the protective clothing to estimate dermal exposure; skin deposition is then extrapolated from patch deposition. Fluorescence imaging involves a simulant that incorporates a fluorescent dye; video imaging of the test individuals provide estimates of dermal deposition. Table 2-1 summarizes the essential features of each technique.

TABLE 2-1 Techniques for Evaluating Protective Clothing

Technique	Measurement	Advantages	Disadvantages
Air sampling	Withdrawal and analysis of air between skin and clothing	Simple analysis	Restricts movement Artificial airflow Provides no skin deposition or absorption data
Passive dosimetry	Analysis of deposition on patches placed on skin or clothing	Ease of sampling Simple analysis	Extrapolation error Provides no skin absorption data
Fluorescence imaging	Measurement of fluorescent tracer deposited on skin	Quantitative assessment of skin deposition	Provides no skin absorption data
Biological monitoring	Measurement of chemical or metabolite in urine, blood, or expired air	Integrates body exposure, skin deposition absorption, and possibly inhalation; Closest estimate of total body dose	Requires information on distribution or metabolism

The biggest advantage of the air sampling technique is the simplicity of direct chemical analysis. However, the removal of air from the space between skin and clothing for sampling purposes can cause local turbulence and create artificial airflow through the fabric. Air lines and connections also restrict the movement of the test individuals. Furthermore, air sampling provides no information on skin deposition and absorption, and biological effects must be extrapolated directly from air concentrations.

The biggest advantage of passive dosimetry is the ease with which samples can be obtained and analyzed. The simulant is usually analyzed, and no metabolism data are necessary. A disadvantage is that deposition on the skin may differ from deposition on the patch; thus no information on real skin deposition is obtained, and skin absorption cannot be calculated. A potential source of error in passive dosimetry is inherent in the extrapolation of deposition from relatively small patches to the entire body surface.

Video imaging of fluorescent tracers is a relatively new technique, which has the advantage of providing a visual image of affected body areas. The technique provides a measure of dermal deposition but assumes that the transport and adsorption of the tracer and chemical simulant are similar. Again, no information on skin absorption is obtained.

The primary advantage of biological monitoring is that the actual internally absorbed dose can be determined provided that information regarding the distribution or metabolism of the compound is available. The technique integrates absorption from all body sites, eliminating the concern over positioning of patches. The integration makes it impossible to pinpoint the location of suit failures; information regarding skin deposition and absorption is inherent in the measurement. Biological monitoring can be done with compounds or their metabolites that are excreted in a variety of biological materials, including blood, urine, feces, expired air, and saliva. The technique has been used successfully by a number of investigators in human studies to determine internally absorbed doses following exposures (Franklin et al., 1981).

Combining two or more testing techniques (e.g., biological monitoring and passive dosimitry) in skin exposure studies has been useful. The first three techniques (air sampling, passive dosimetry, fluorescence imaging) are best suited to determining exposure (i.e., the amount of chemical that reaches the skin). These techniques would be most helpful for designing effective gear. In contrast, biological monitoring determines an absorbed dose and would be most useful for assessing a soldier's effectiveness.

MIST Test Procedure

The MIST and BRHA were developed to provide a system-level evaluation of CB protective suits (Fedele and Nelson, 1996). Briefly, each subject in a group of up to eight volunteers is fitted with approximately 20 passive detector patches for methyl salicylate vapor, which simulates the vapors of VX (an organophosphorus cholinesterase inhibitor) or HD (the vesicant bis(2-chloroethyl) sulfide). The detectors, which contain an absorbent powder (Tenax TA), are placed at various anatomical locations (Figure 2-1). The subjects don test suits, gas masks, and hoods and enter a test chamber containing methyl salicylate at a concentration of 100 mg/m^3 at an air temperature of 21°C to 32° C, 50 to 80 percent relative humidity, and an airflow rate of 3 to 16 kph (2 to 10 mph).

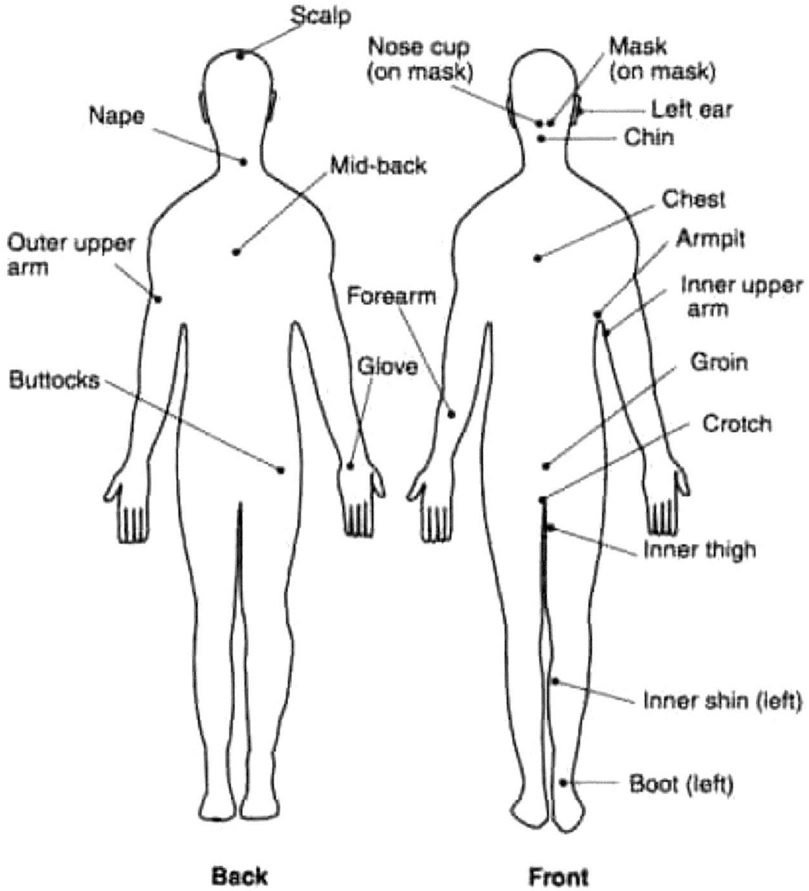

Figure 2-1 Placement of passive sampling devices for the MIST.

The test chamber is a stainless steel, environmentally-controlled unit that can accommodate eight test subjects. The chamber is maintained at a negative pressure by an air filtration system and a controlled air intake system. The chamber was designed for agent testing of large systems and has a design specification to provide controlled temperature (-32°C to 38°C) and a nominal wind speed of 2 to 10 mph, provided by two large propellers to ensure a unidirectional airflow, which is recirculated through the chamber through a false ceiling and a series of vents (see Figure 2-2).

The test chamber is supported by several modules: a control room on the west side of the chamber, which is the central location for monitoring and controlling all physical parameters; two separate instrumentation rooms; and an airlock chamber entry/exit room, which allows test participants to enter and exit the chamber without releasing large amounts of simulant to the atmosphere. The egress area is also used as a clean undress location (Hanzalka et al., 1996).

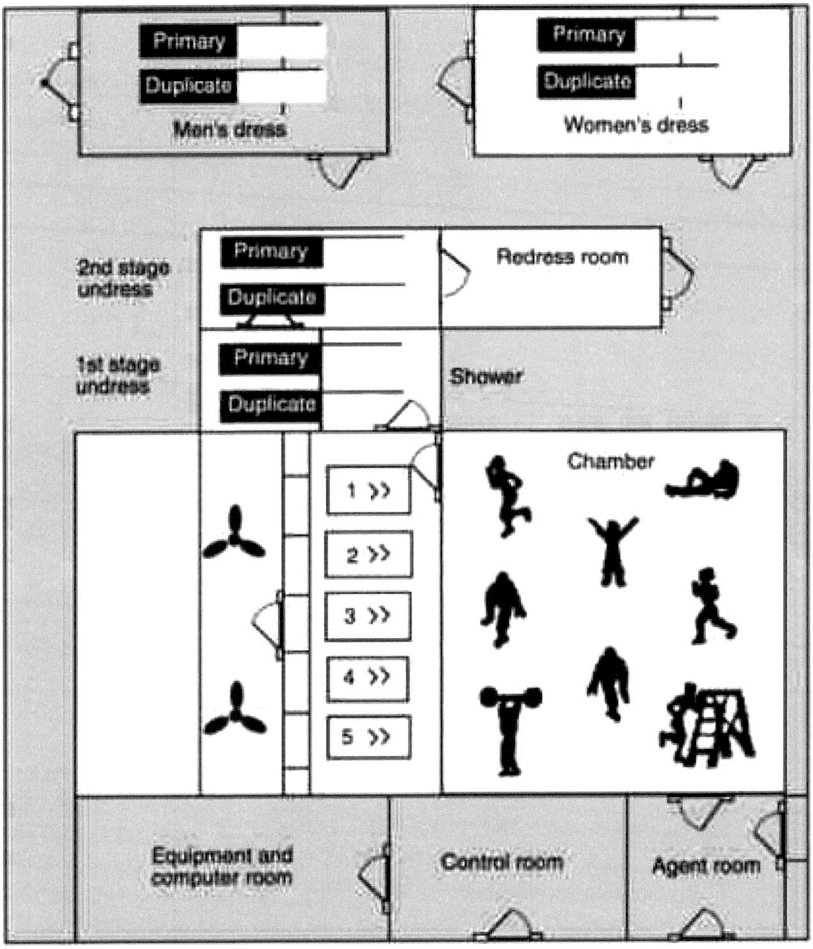

Figure 2-2 MIST system test chamber.

Additional subjects wear suits virtually impermeable to vapor (i.e., control suits) with corresponding detectors on the outer surface of the suit. All subjects follow a prescribed exercise or movement routine for the duration of the exposure (120 min). The exercise protocol consists of eight "stations." Station #1 consists of jumping jacks, Station #2 seated rest, Station #3 walking simulation (treadmill), Station #4 seated rest, Station #5 moving weights, Station #6 a "take cover" maneuver, Station #7 walking simulation (treadmill), and Station #8 climbing and reaching (Hanzelka et al., 1996). The threat or challenge is defined in terms of a concentration x time factor (Ct) of 12,000 mg/m^3-min (100 mg/m^3 x 120 min). After the subjects exit the test chamber, the detectors are removed for gas chromatographic analysis. Because the background level of methyl salicylate in control detectors ranges from 50 to 100 nanogram (ng), detectors with values of less than 100 ng are assigned a value of 100 ng.

Body Region Hazard Analysis

When the methyl salicylate is intended to simulate mustard vapor (HD), the BRHA is "local" because HD exerts its toxicity primarily on localized regions of the skin. When exposure to VX vapor is being simulated, the BRHA is "systemic" because VX toxicity results from cumulative absorption through exposed skin.

The mass of VX required to produce systemic toxicity after exposure to various regions of the skin was derived from the work of Sim (1962) who estimated the dose of liquid VX required to cause a 70 percent depression in cholinesterase from studies in which droplets of VX were applied to various skin sites on volunteer subjects (see Table 2-2 and the Acetylcholinesterase Inhibition as a Biological Marker in Chapter 5). The symptoms exhibited by the subjects included local sweating, erythema, weakness, muscular fasciculation, dizziness, headache, abdominal cramps, repeated vomiting, and diarrhea. The wide range of the data in Table 2-2 reflects the variation of skin permeability at different anatomic sites. Factors that affect the regional permeability of the skin have not been precisely defined. However, the thickness of the stratum corneum and the density of adnexal structures (e.g., hair follicles) may be contributors. VX does not appear to undergo appreciable metabolism by the skin, in contrast to soman, which undergoes hydrolysis (van Hooidonk et al., 1983). At a given anatomic site, the absorption rate generally increases as the dose increases, whereas the efficacy of absorption (the percentage of agent absorbed) generally decreases as the dose increases. The effects of dosage on regional variations in skin permeability have not been adequately studied.

TABLE 2-2 Estimated Topical Dose of VX That Would Cause a 70 Percent Depression in Red Blood Cell Cholinesterase in a 70-kg Human

Area of Application	Single-Drop Dose of VX (mg)
Cheek	0.36
Ear	0.46
Top of head	0.76
Forehead	0.78
Groin	1.22
Back of neck	1.72
Axilla	2.07
Popliteal space	2.09
Abdomen	2.23
Elbow	2.25
Back	2.65
Forearm (volar)	2.80
Hand (dorsum)	2.91
Buttocks	4.26
Forearm (dorsum)	6.57
Foot (dorsum)	6.60
Foot (plantar)	7.14
Knee	7.14
Hand (palmar)	9.24

The BRHA assumes that (1) the effective vapor exposure (*Ct* factor) for each region of the body would produce toxicity (e.g., nausea and vomiting); in other words, when vapor is presented to that body region only, toxicity will result; and (2) the relative differences in exposures to VX vapor required to produce an equivalent toxicity from the various skin sites are the same as the corresponding differences for the exposures to VX liquid.

For exposure to HD vapor, the *Ct* factor necessary to produce equivalent toxicity at various anatomic locations is based on a vapor exposure of 1,000 mg/m^3-min to a human forearm causing severe burns in hot (above 27°C), humid conditions. The BRHA calculates the exposure necessary to produce equivalent toxicity at other anatomic locations based on the regional toxicity data for VX. For example, the amount of VX required to cause equivalent toxicity after application in the chin and neck area (5.1 μg/kg, Table 2-3) is approximately one-eighth of the corresponding amount for the forearm (40 μg/kg). The associated HD vapor exposure for chin and neck is then calculated as approximately one-eighth of the forearm vapor. As indicated in Table 2-3 and by other data, the scrotum and areas of the head and neck are particularly vulnerable because of higher skin permeability. These areas are also near discontinuities in the typical chemical protective ensemble (CPE), which increases their vulnerability. These areas should receive special attention in the design of the new CPE.

Evaluation of Suits against Exposure to HD (Local Analysis)

To calculate the "protection value" of test suits against a simulated exposure to HD vapor (local analysis), the "protection factor" for each anatomical site is calculated by dividing the amount of methyl salicylate found in the detector on the outside of the control suit by the amount found in the corresponding detector on the inside of the test suit. Each site protection factor is multiplied by the exposure to HD necessary to produce severe bums (column 4 in Table 2-3). The lowest value of the resulting site protection factors is assigned as the test suit's protection value, expressed as a *Ct* value in mg/m^3-min. The results for tests of 41 different candidate ensembles are given in Table 2-4, which gives the number of test replicates (*n*) and the geometric mean and standard deviation of the local effect of concentration vs. time (*Ct*) for each protective ensemble. Ensembles are ranked from most protective (1) to least protective (41). The battle dress uniform (41) is a standard military uniform not specifically designed for chemical protection. Table 2-5 contains the log_e transform of the geometric mean and standard deviation contained in Table 2-4, as well as the 95 percent confidence intervals (95 percent CIs) for the log_e (local effective *Ct*).

The geometric: mean[1] has been used to average ratios (here protection factors) when each ratio is to be given equal weight. A log transform of local or systemic *Ct* has the effect of further reducing the

[1] The geometric mean, the nth root of the product of the *n* data or $(\sqrt[n]{x_1 x_2 x_3 \ldots x_n})$ was used in the calculation to give each test result equal weight (Zar, 1984).

variability of the data. An alternate statistical analysis of the data should be done to investigate the influence of these mathematical treatments on the ranking of CPEs. The data should also be analyzed for the influence of test subject characteristics, such as body size and weight.

TABLE 2-3 Parameters for Local Body Region Hazard Analysis[a]

Region	Area (cm^2)	VX Whole Body Dose (μg/kg)	Local Exposure to HD (mg/m^3-min)
Scrotum	200	1.6	39
Chin and Neck	200	5.1	1290
Ears	50	6.6	164
Cheeks and neck	100	6.8	171
Nape (back of neck)	100	24.6	614
Scalp (top of head)	350	10.8	271
Abdomen	2,858	31.8	796
Back	2,540	37.9	946
Buttocks	953	60.9	1,521
Arms (lower, volar)	487	40.0	1,000
Arms (upper, volar)	488	40.0	1,000
Elbows (back)	50	32.2	804
Arms (lower, dorsum)	706	93.8	2,346
Arms (upper, dorsum)	706	93.8	2,346
Legs (plantar, lower)	948	40.0	1,000
Legs (plantar, upper)	1,422	60.9	1,521
Legs (dorsum, lower)	1,897	93.8	2,346
Legs (dorsum, upper)	2,845	93.8	2,346
Knees (front)	200	102	2,550

[a] Anatomical site, surface area of anatomical site, dose of VX that would cause a 70 percent depression in red blood cell cholinesterase, and estimated local exposure to HD (mg/m^3-min) that would cause severe burns.

Source: Fedele and Nelson (1996).

TABLE 2-4 Ranking of Protective Ensembles by Local Effective Ct[a]

Rank	Ensemble[b]	N[c]	Geometric Mean[d]	Geometric Standard Deviation[e]
1	Duty uniform, non-fire resistant, model 1	8	8,730	2.44
2	Duty uniform, fire resistant, model 1 (w)	2	7,614	1.39
3	Overgarment, non-fire resistant, model 3 (w)	6	6,933	1.67
4	Saratoga chemical protective suit (worn over personal undergarments only or over duty uniform) (w)	6	5,800	1.78
5	Duty uniform, non-fire resistant, model 3	7	5,689	2.86
6	Duty uniform, non-fire resistant, model 2	4	5,317	1.89
7	Overgarment, non-fire resistant, model 2 (r)	6	5,079	2.31
8	Duty uniform, fire resistant, model 2	10	4,331	2.04
9	Overgarment, non-fire resistant, model 2 (w)	6	5,079	3.33
10	Overgarment, fire resistant, model 2 (w)	6	4,096	3.86
11	Duty uniform, fire resistant, model 1	6	4,052	2.38
12	Overgarment, fire resistant, model 1 (w)	6	3,283	3.36
13	Overgarment, non-fire resistant, model 2	23	3,089	2.88
14	Duty uniform, non-fire resistant, model 3 (w)	6	2,835	1.99
15	Overgarment, non-fire resistant, model 1 (r)	6	2,778	1.88
16	Overgarment, fire resistant, model 2	23	2,700	2.95
17	Vapor protective,fire resistant, protective undergarment	14	2,492	3.84
18	Overgarment, non-fire resistant, model 1	19	2,441	3.52
19	Overgarment, fire-resistant, model 1	29	2,308	3.01
20	Overgarment, non-fire resistant, model 2 (r)	6	2,129	2.12
21	Vapor protective, fire resistant, protective undergarment (w)	4	2,101	2.08
22	Duty uniform, fire resistant, model 2 (w)	6	2,002	2.33
23	Standard U.S. Army chemical protective undergarment	24	2,001	2.25
24	Duty uniform, non-fire resistant, model 2 (w)	6	1,930	2.08
25	Overgarment, non-fire resistant, model 3	19	1,899	3.67
26	Duty uniform, non-fire resistant, model 1 (w)	6	1,823	3.82
27	Saratoga chemical protective suit (worn over personal undergarments only or over duty uniform)	15	1,804	2.25
28	Standard Army chemical protective undergarment (w)	4	1,305	1.97
29	Battle dress overgarment	29	1,295	2.35
30	Overgarment, non-fire resistant, model 1 (w)	6	1,221	1.82
31	Overgarment, non-fire resistant, model 2 (nr)	4	1,124	1.72
32	Air crew uniforms, model 2	15	1,045	2.42
33	Air crew uniforms, model 1	10	930	2.07
34	Air crew uniforms, model 1 (w)	6	846	1.98
35	Overgarment, non-fire resistant, model 1 (nr)	4	680	2.24
36	Air crew uniforms, model 2 (w)	6	594	1.28
37	Overgarment, non-fire resistant, model 3 (nr)	4	527	1.84
38	Saratoga chemical protective suit (worn over personal undergarments only or over duty uniform (r)	5	515	1.57
39	Standard U.S. Army chemical protective undergarment	2	447	1.14
40	Saratoga chemical protective suit (worn over personal undergarments only or over duty uniform) (nr)	5	423	2.33
41	Battle dress uniform (standard military uniform)	4	53	1.19

[a] Ct is agent concentration (mg/m^3) x time of exposure (in minutes)
[b] Various protective ensembles tested (w = worn, r = repaired, nr = not repaired)
[c] Number of test replicates for each ensembled
[d] The geometric mean, the nth root of the product of the n data or $\sqrt[n]{x_1 x_2 x_3 \ldots x_n}$
[e] The geometric standard deviation, Σ (standard deviation of logarithms of values)

TABLE 2-5 Analysis of Variance (ANOVA) on *Log*$_e$ (Local Effective *Ct*) of Protective Ensembles

		Log$_e$ (local effective *Ct*)		
Rank	N[a]	*Log*$_e$ (Geometric Mean)	*Log*$_e$ (Geometric Standard Deviation)	Based on Pooled Standard Deviation[b]
1	8	9.1	0.9	(- - • - -)
2	2	8.9	0.3	(- - - - - - • - - -)
3	6	8.8	0.5	(- - - • - - -)
4	6	8.7	0.6	(- - • - - -)
5	7	8.7	1.0	(- - • - - -)
6	4	8.6	0.6	(- - - - • - - - -)
7	6	8.5	0.8	(- - - • - -)
8	10	8.4	0.7	(- - • - -)
9	6	8.3	1.2	(- - - • - - -)
10	6	8.3	1.4	(- - - • - -)
11	6	8.3	0.9	(- - - • - - -)
12	6	8.1	1.2	(- - • - - -)
13	23	8.0	1.1	(- • -)
14	6	8.0	0.7	(- - - • - - -)
15	6	7.9	0.6	(- - - • - -)
16	23	7.9	1.1	(- •)
17	14	7.8	1.0	(- • - -)
18	19	7.8	1.3	(- • -)
19	29	7.7	1.1	(- •)
20	6	7.7	0.8	(- - • - - -)
21	4	7.6	0.7	(- - - • - - - -)
22	6	7.6	0.8	(- - - • - - -)
23	24	7.6	0.8	(- • -)
24	6	7.6	0.7	(- - - • - - -)
25	19	7.6	1.3	(- • -)
26	6	7.5	1.3	(- - - • - -)
27	15	7.5	0.8	(- • - -)
28	4	7.2	0.7	(- - - - • - - - -)
29	29	7.2	0.9	(- • -)
30	6	7.1	0.6	(- - - • - -)
31	4	7.0	0.5	(- - - - • - - - -)
32	15[a]	7.0	0.9	(- - • -)
33	10	6.8	0.7	(- - • - -)
34	6	6.7	0.7	(- - - • - - -)
35	4	6.5	0.8	(- - - - • - - -)
36	6	6.4	0.2	(- - - • - - -)
37	4	6.3	0.6	(- - - • - - - -)
38	5	6.2	0.4	(- - - • - - -)
39	2	6.1	0.1	(- - - - - - • - - - - -)
40	5	6.0	0.8	(- - - • - - -)
41	4	4.0	0.2	(- - - - • - - - -)
				4.0 6.0 8.0 10.0

[a] Number of test replicates for each ensemble

[b] Pooled standard deviation = 1.0

Evaluation of Suits against Exposure to VX (Systemic Analysis)

To calculate the protection value of test suits against a simulated exposure to VX vapor (systemic analysis), the protection factor for each anatomical site is calculated in the same manner as for the local analysis. The site protection factors (PF_{site}) along with the site size ($skin\ area_{site}$) and the VX dose site ($VX\ dose_{site}$) that would cause a 70 percent depression of red blood cell cholinesterase (Table 2-2) are used to calculate a "whole body effective exposure" (WBEE) according to the following formula:[2] where the total body surface area (excluding the face) is 18,950 cm^2. This calculation is essentially the sum of the VX dose sites that have been weighted for the area and protection factor of each site. The WBEE for an unprotected person weighing 70-kg was calculated by setting all the site protection factors equal to unity to yield a value of 2.45 mg for the data in Table 2-3.

$$WBEE\ (mg) = \frac{18{,}950\ (cm^2)}{SUM\ [\ (skin\ area_{site})/(PF_{site})(VX\ dose_{site})\ ]}$$

The protection factor for a test suit is calculated by dividing the WBEE (mg) by 2.45 mg. The protection factor for a test suit is multiplied by the *Ct* factor for VX (25 mg/m^3-min), which is assumed to be the vapor equivalent of the WBEE for the unprotected person (Reutter and Wade, 1994). This figure is called the "systemic effective *Ct*," and the geometric mean of replicate evaluations is used to rank the test suits relative to the battle dress overgarment. The results from tests involving 41 protective suits are given in Table 2-6. Geometric mean, geometric standard deviation, and CIs for these data are given in Table 2-7.

Table 2-8 shows the total mass of methyl salicylate collected in the passive samplers and the geometric mean of the local effective *Ct* (local analysis for HD) for the 41 protective ensembles (see also Figure 2-3). There is no linear relationship between total mass and local effective *Ct* for the protective ensembles. Instead, there may be an exponential relationship of the type [$Y = A^{(-kx)}$] where Y is the local effect *Ct* and x is the total sampler mass. This equation is plotted in Figure 2-3, where the constants are A = 14,600 and k= 3.40 x 10^{-5}. The results suggest that the total sampler mass alone may be useful for the preliminary ranking of chemical protective values.

---

[2] See Appendix A for the derivation of this equation.

TABLE 2-6 Ranking of Protective Ensembles by Systemic Effective Ct^a

Rank	Ensemble[b]	N[c]	Geometric Mean[d]	Geometric Standard Deviation[e]
1	Vapor protective, fire resistant, protective undergarment	14	3,225	2.12
2	Duty uniform, non-fire resistant, model 1	8	3,177	2.24
3	Duty uniform, non-fire resistant, model 3	7	3,003	1.89
4	Duty uniform, non-fire resistant, model 2	4	2,466	1.56
5	Duty uniform, fire resistant, model 1	6	2,463	2.05
6	Overgarment, non-fire resistant, model 3 (r)	6	2,409	1.57
7	Overgarment, non-fire resistant, model 3 (w)	6	2,278	1.30
8	Overgarment, non-fire resistant, model 2 (w)	6	2,260	1.76
9	Saratoga chemical protective suit (worn over personal undergarments only or over duty uniform) (w)	6	2,230	1.24
10	Vapor protective, fire resistant, protective undergarment (w)	4	2,189	1.80
11	Duty uniform, non-fire resistant, model 2	10	2,122	1.90
12	Overgarment, fire-resistant, model 2 (w)	6	2,033	1.92
13	Standard U.S. Army chemical protective undergarment	24	1,835	1.76
14	Overgarment, non-fire resistant, model 2	23	1,759	1.80
15	Duty uniform, non-fire resistant, model 3 (w)	6	1,725	1.49
16	Duty uniform, fire resistant, model 2 (w)	6	1,701	1.97
17	Overgarment, non-fire resistant, model 1 (w)	6	1,698	2.40
18	Overgarment, fire resistant, model 2	23	1,679	1.83
19	Overgarment, non-fire resistant, model 1	19	1,612	1.80
20	Overgarment, fire resistant, model 1	29	1,552	2.05
21	Saratoga chemical protective suit (worn over personal undergarments only or over duty uniform)	15	1,552	1.66
22	Standard U.S. Army chemical protective undergarment (w)	4	1,540	1.50
23	Duty uniform, fire resistant, model 2 (w)	6	1,531	1.67
24	Duty uniform, fire resistant, model 1 (w)	2	1,506	1.32
25	Battle dress overgarment	29	1,494	1.91
26	Overgarment, non-fire resistant, model 1(r)	6	1,285	1.27
27	Overgarment, fire resistant, model 3	19	1,201	2.08
28	Overgarment, non-fire resistant, model 2 (r)	6	1,124	1.37
29	Duty uniform, non-fire resistant, model 1 (w)	6	972	2.65
30	Air crew uniforms, model 2	15	944	1.49
31	Air crew uniforms, model 1 (w)	6	885	1.99
32	Air crew uniforms, model 1	10	857	1.55
33	Air crew uniforms, model 2 (w)	6	838	1.27
34	Overgarment, non-fire resistant, model 1 (w)	6	716	1.67
35	Standard U.S. Army chemical protective undergarment	2	585	1.35
36	Saratoga chemical protective suit (worn over personal undergarments only or over duty uniform (r)	5	487	1.64
37	Overgarment, non-fire resistant, model 3 (nr)	4	451	1.53
38	Overgarment, non-fire resistant, model 1 (nr)	4	429	1.32
39	Overgarment, non-fire resistant, model 2 (nr)	4	411	1.07
40	Saratoga chemical protective suit (worn over personal undergarments only or over duty uniform (nr)	5	270	1.39
41	Battle dress uniform (standard military uniform)	4	38	1.07

[a] Ct is agent concentration (mg/m^3) x time of exposure (in minutes)
[b] Various protective ensembles tested (w = worn, r = repaired, nr = not repaired)
[c] Number of test replicates for each ensemble
[d] The geometric mean, the nth root of the product of the n data or $\sqrt[n]{x_1 x_2 x_3 \ldots x_n}$
[e] The geometric standard deviation, Σ (standard deviation of logarithms of values)

TABLE 2-7 Analysis of Variance (ANOVA) on *Log*$_e$ (Systemic *Ct*) of Protective Ensembles

		Log$_e$ (local effective *Ct*)		
Rank	N[a]	*Log*$_e$ (Geometric Mean)	*Log*$_e$ (Geometric Standard Deviation)	Based on Pooled Standard Deviation[b]
1	14	8.1	0.8	(-•-)
2	8	8.1	0.8	(--•--)
3	7	8.0	0.6	(--•--)
4	4	7.8	0.4	(---•---)
5	6	7.8	0.7	(--•--)
6	6	7.8	0.5	(--•--)
7	6	7.7	0.3	(---•--)
8	6	7.7	0.6	(--•---)
9	6	7.7	0.2	(--•---)
10	4	7.7	0.6	(---•---)
11	10	7.7	0.6	(-•--)
12	6	7.6	0.6	(--•--)
13	24	7.5	0.6	(•-)
14	23	7.5	0.6	(-•)
15	6	7.4	0.4	(---•--)
16	6	7.4	0.7	(---•--)
17	6	7.4	0.9	(---•--)
18	23	7.4	0.6	(-•)
19	19	7.4	0.6	(•-)
20	29	7.4	0.7	(•)
21	15	7.4	0.5	(-•-)
22	4	7.3	0.4	(---•---)
23	6	7.3	0.5	(--•--)
24	2	7.3	0.2	(-----•----)
25	29	7.3	0.6	(-•)
26	6	7.2	0.2	(--•--)
27	19	7.1	0.7	(-•-)
28	6	7.0	0.3	(--•--)
29	6	6.9	1.0	(--•--)
30	15	6.8	0.4	(-•-)
31	6	6.8	0.7	(--•--)
32	10	6.8	0.4	(-•-)
33	6	6.7	0.2	(--•--)
34	6	6.6	0.5	(--•--)
35	2	6.4	0.3	(----•-----)
36	5	6.2	0.5	(--•---)
37	4	6.1	0.4	(---•---)
38	4	6.1	0.3	(---•---)
39	4	6.0	0.1	(---•---)
40	5	5.6	0.3	(---•---)
41	4	3.6	0.1	(----•----)
				4.5 6.0 7.5

[a]Number of test replicates for each ensemble
[b]Pooled standard deviation = 0.6

TABLE 2-8 Comparison of Total Mass of Methyl Salicylate Collected in Passive Samplers versus Geometric Mean of Local Effective *Ct* (Local Analysis for Mustard Gas)

Protective Ensembles[a]	Total Mass (μg) Collected in Samplers	Geometric Mean of Local Effective *Ct* (mg/m^3-min)
Duty uniform, fire resistant, model 1	24.2	8,730
Duty, non-fire resistant, model 3	25.1	5,690
Overgarment, non-fire resistant, model 3 (w)	27.5	6,930
Duty uniform, non-fire resistant, model 2	28.7	5,320
Overgarment, non-fire resistant, model 3 (r)	32.1	5,080
Vapor protective, fire resistant, protective undergarment	32.4	2,490
Duty uniform, fire resistant, model 1	33.6	4,050
Saratoga chemical protective suit (worn over personal undergarments only or over duty uniform (w)	33.8	5,800
Overgarment, non-fire resistant, model 2 (w)	33.9	4,210
Overgarment, fire resistant, model 2 (w)	36.5	4,100
Duty uniform, non-fire resistant, model 3 (w)	37.4	2,840
Duty uniform, fire resistant, model 2 (w)	38.0	2,000
Duty uniform, fire resistant, model 2	39.6	4,330
Vapor protective, fire resistant, protective undergarment (w)	40.0	2,100
Duty uniform, fire resistant, model 1 (w)	42.0	7,610
Standard U.S. Army chemical protective undergarment	47.7	2,000
Overgarment, fire resistant, model 2	48.0	2,700
Saratoga chemical protective suit (worn over personal undergarments only or over duty uniform)	49.1	1,800
Overgarment, non-fire resistant, model 2	49.3	3,090
Overgarment, fire resistant, model 1 (w)	49.3	3,280
Overgarment, non-fire resistant, model 1 (r)	49.9	2,780
Duty uniform, non-fire resistant, model 2 (w)	51.3	1,930
Overgarment, fire resistant, model 1	55.6	2,310
Overgarment, non-fire resistant, model 2 (r)	56.6	2,130
Standard U.S. Army chemical protective undergarment (w)	57.4	1,300
Overgarment, non-fire resistant, model 1	58.0	2,440
Battle dress overgarment	71.2	1,300
Overgarment, non-fire resistant, model 3	73.8	1,900
Air crew uniforms, model 2	94.8	1,040
Duty uniform, non-fire resistant, model 1 (w)	99.6	1,820
Air crew uniforms, model 2 (w)	102.4	590
Air crew uniforms, model 1	103.8	930
Overgarment, non-fire resistant, model 1 (w)	107.2	1,220
Overgarment, non-fire resistant, model 3 (nr)	114.9	530
Air crew uniforms, model 1 (w)	115.3	850
Saratoga chemical protective suit (worn over personal undergarments only or over duty uniform (r)	115.5	510
Overgarment, non-fire resistant 1 (nr)	135.5	680
Standard U.S. Army chemical protective undergarment	151.3	450
Overgarment, non-fire resistant, model 2 (nr)	156.6	1,120
Saratoga chemical protective suit (worn over personal undergarments only or over duty uniform (nr)	209.0	420
Battle dress uniform, (standard military uniform)	1,788.8	50

[a]w = worn, r = repaired, nr = not repaired

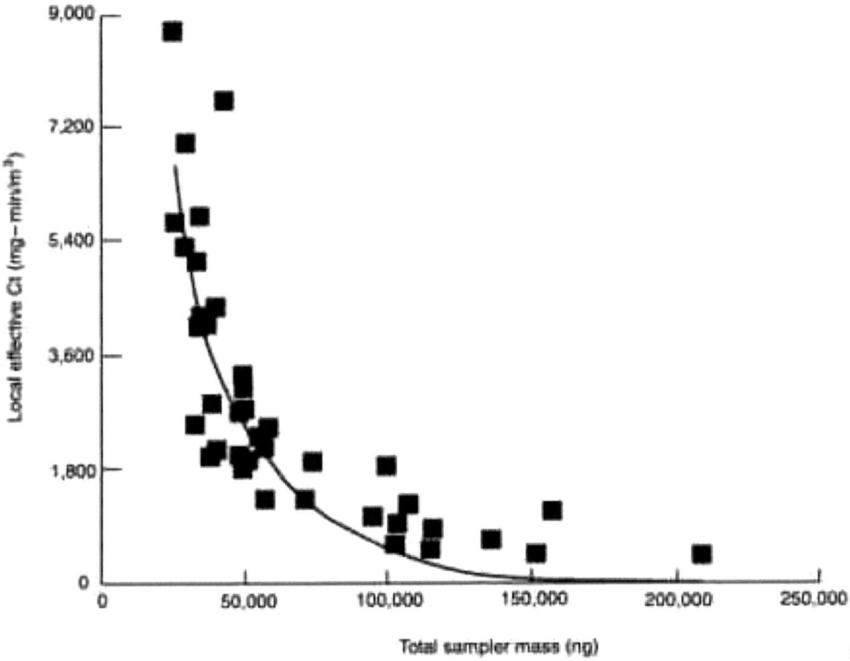

Figure 2-3 Local effect *Ct* versus total sampler mass.

3

Simulant Section

IMPORTANCE OF DERMAL PENETRATION

The importance of protective ensembles to prevent penetration of vapors is based on the consequence of dermal exposure. Dermal penetration has been shown to play a significant role in exposure to both chlorinated pesticides and organophosphates. In an attempt to account for both dermal penetration and inhalation, Finland and the United States now monitor chlorophenols in urine for setting air standards for workers exposed to chlorinated compounds. The importance of dermal exposure has also been shown in the treatment of wood (Fenske et al., 1987; Kauppinen and Lindross, 1985). Dermal exposure to organophosphates has also been found to be significant (Environmental Protection Agency, 1992).

The effects of changes in the skin barrier can be critical. For example, although parquet is not known to be absorbed by the skin, a fatal case involving skin penetration has been reported (Newhouse et al., 1978). In this case, the patient had numerous scratches on his arms and legs, and skin absorption over a period of time was fatal. Besides cuts and scratches, other conditions that enhance dermal penetration include skin hydration and dermatitis. Fenske and co-workers (Fenske et al., 1987) reported on skin contamination by tetrachlorophenol (detected by a fluorescent tracer) in timber workers who wore polyvinyl chloride gloves. More than 86 percent of the contamination was detected on the palms (in the case of one worker, a cut through the glove material was found). As all of these and other studies have shown, dermal penetration by chemical agents can be significant. Consequently, evaluating the potential penetration of protective ensembles is necessary for determining the amount of chemical that might be deposited on the skin and/or absorbed through the skin.

Because of the acute toxicity of the agents being considered, simulant compounds must be used for testing. Regardless of the permeability of the simulant, passive dosimitry is an effective testing method for determining the relative distribution of the chemical to assess potential problem areas of the suit. Most problem areas appear to be at closures or around tears. If a simulant is used to predict actual skin penetration, the simulant should have similar physical and chemical characteristics as the agent of concern. Otherwise, comparative absorption rates, specific to each body region for each agent and simulant, must be developed.

USE OF SIMULANT TO PREDICT DERMAL PENETRATION

Protective garments may be required for an array of chemical agents. These include the organophosphate nerve agents, GA, GB, and VX, as well as the vesicant blister formulations of sulfur mustard, H, HD, and HT. Protection against these agents may be needed during wars, during terrorist attacks (such as the Japanese subway incident).

The World Health Organization (WHO) (1990), has reported that each year approximately 3 million people worldwide are poisoned by pesticides (nerve agents) resulting in 220,000 deaths. (According to the WHO, acute poisonings, including suicide attempts, mass poisonings from contaminated food, chemical accidents in industry, and occupational exposure in agriculture constitute the most serious health hazards from agricultural pesticides). Operations like Desert Shield and Desert Storm revealed the need for protective clothing. Because of the aggressive toxicity of nerve agents, the highest quality of protection is critical. At the same time, the soldier's operational capability must not be impaired.

In the 1940s, mustard gas was used on human test subjects. Observations showed that infiltration occurred at the head, neck, and ankles. The suits that were tested performed well except for closure areas. However, testing on human subjects with real agents has been discontinued. Therefore, an appropriate surrogate compound must now be used to test protective clothing.

There are two basic types of surrogate compounds for chemical warfare agents (CWAs): analogs and simulants. Analogs are not classified as CWAs but are structurally similar and are considerably toxic. Examples are chloroethyl ethyl sulfide, a powerful vesicant, and diisopropyl fluorophosphonate, a potent cholinesterase inhibitor.

Simulants have similar physical properties to CWAs but are essentially nontoxic. Simulants for fabric testing include methyl salicylate (MeS) for HD, dimethyl methylphosphonate for VX, and di-isopropyl methyl phosphonate for GB. Challenges from a surrogate should reveal weaknesses in protective gear. For example, in a 1990 test evaluating vapor protection capabilities of the jacket and trouser interface of a chemical protective ensemble, the mean vapor level measured in the abdominal area was 17 percent of the outside level. This demonstrated that a significant amount of vapor infiltrated the CPE through the interfaces and could pose a threat to the wearer (Scott and Pointer, 1990).

The military has used simulants of chemical agents in a variety of studies where it was important to estimate the disposition or movement of chemical agents and where a toxic endpoint was either unnecessary or undesirable. A simulant was selected on the basis of low toxicity and similarity of certain physical properties to the chemical agent. Chemical structure could be quite different. For example, MeS has a structure that bears no similarity to HD or VX (Figure 3-1), but its vapor pressure, density, and water solubility are similar to those of HD (Table 3-1). The general assumption is that chemicals with similar chemical and physical properties will behave in similar ways. For example, it is known that at low to moderate pressures, binary diffusion coefficients vary inversely with pressure or density (Reid et al., 1987). These coefficients would be important for selecting chemical agent simulants for MIST.

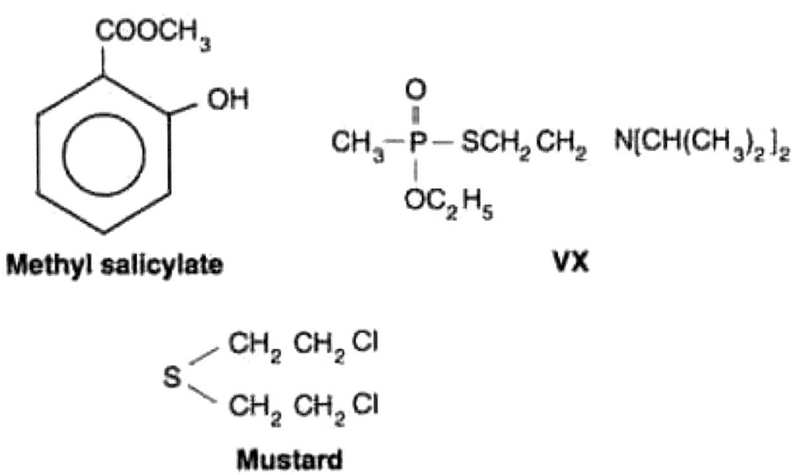

Figure 3-1 Chemical structures of methyl salicylate, VX, and mustard.

[a]

Chemical Agent	GB	VX	H, HD	HT	MeS
Common Name	Sarin	-	Sulfur mustard	Sulfur mustard	Methyl salicylate
CAS No.	107-44-8	50782-69-9	505-60-2	Blend	119-36-8
Chemical Formula	$C_4H_{10}FO_2P$	$C_{11}H_{26}NO_2PS$	$C_4H_8Cl_2S$	Blend	$C_80_3H_8$
Molecular weight	-	267	159 (HD)	-	152
Vapor Pressure (@ 25°C mm Hg)	2.9	0.0007	0.08	0.104	0.091[c] (20° C)
Liquid Density (@ 25°C g/cm³)	1.089	1.008	1.27	1.27	1.18[b]
Freezing Point (°C)	-56	-39	8–12(H)	-	-8.3[c]
Water Solubility (g/m @25° C)	∝[d]	3.1[d]	0.09[d]	-	0.07[e]
Mode of Action	Nerve agent	Nerve agent	Blister agent	Blister agent	Relatively nontoxic

[a] Characteristics of agents from Daughtery et al., 1992.
[b] Conkle et al., 1986.
[c] Arca, 1996.
[d] Reifenrath, 1980.
[e] The Merck Index, 1996.

A gas can move through protective clothing by sorption onto the ensemble surface, diffusion into the material, and desorption of the molecules from the inner surface of the fabric. A gas can also move through closures, seams, and imperfections in protective clothing. Component level tests of fabric swatches showed that penetration by chemical agents and MeS occurred only after three or four days of exposure (Dugway Proving Ground, 1994). Therefore, MeS penetration in the two hour system level MIST test was probably due to movement of the gas through discontinuities in the ensembles. It follows that the value of the MIST conducted with MeS was in predicting ensemble fit and closure integrity.

SELECTION OF A SIMULANT

A number of simulant compounds have been used to predict agent penetration of ensembles. In 1994, the Army attempted to determine the relationship between fabric penetration by the vesicant agent HD and the simulant MeS in order to correlate the penetration of these two chemicals. The results of those tests were the basis for choosing MeS as a simulant for the MIST program (Dugway Proving Ground, 1994).

The Army has conducted penetration tests of various fabrics with both HD and MeS vapors. Vapor breakthrough was evaluated by plotting penetration curves (breakthrough concentration vs. cumulative *Ct*). *Ct* accounts for minor fluctuations in concentration during the test and represents the true loading of agent on fabric.

The resulting penetration curves indicated that MeS penetrates the fabrics about 30 percent slower than HD. It appears that the initial breakthrough occurs after three or four days of challenge (Dugway Proving Ground, 1994). Breakthrough levels increase gradually to about 5 percent, after which there is a dramatic increase. The data reveals a large amount of scatter. However, breakthrough curves were similar enough to support MeS as a simulant for evaluating the ensembles to protect against HD. Some similarities between the physical and chemical characteristics of MeS and other agents suggest that MeS may also be an appropriate surrogate for organophosphates. However, the significant difference in vapor pressure between MeS and nerve agents suggests that MeS would not be as good a simulant for nerve agents (see Table 3-1).

Because vapor penetration of the fabric was reported only after three or four days after onset of challenge, the two-hour testing period can be expected to test only penetration through seams and closure areas. It is not possible to conclude from the MIST that the most significant penetration of CPEs in actual use occurs around closures because the MIST results would logically be skewed toward penetration through closure areas. The MIST can, however, predict closure/seam areas of greatest penetration. This information is valuable for the immediate protection of personnel under attack.

Because penetration during the first few days of exposure are most significant at closure regions, any simulant might be able to provide the necessary information. In this case, both closure areas and folds could change because test subjects don ensembles slightly differently on different days. The mean suit penetration is the most important measurement for comparing garments.

The similar physical properties of MeS and HD support the use of MeS in the MIST. However, MeS and the nerve agents (satin and VX) do not have similar physical properties. Additional data would be required to support the use of MeS as a simulant for measuring the skin penetration of blister or nerve agents.

TOXICITY OF METHYL SALICYLATE

MeS is an oily liquid with wintergreen flavor. It is used in perfumes; as a flavoring in foods, beverages, and pharmaceuticals; and as an ultraviolet absorber in sunburn lotions. MeS is not acutely toxic (lethal doses are reported as 30 ml for adults and 10 ml for children) and does not cause dermal irritation in humans (Sax and Lewis, 1987). It has been used historically as a simulant and has been approved for use with humans by the Army's surgeon general. The environmental toxicity and persistence of MeS have also been studied (Cataldo et al., 1994), including the interaction of MeS with foliage and soils.

FINAL CHOICE OF A SIMULANT FOR MIST

The Army's choice of MeS as a simulant was based on the following information:

- Tests for MeS and HD have been conducted with protective suit fabrics. Penetration curves indicate that MeS penetrates fabrics about 30 percent slower than HD, but the differences in penetration rates are not as significant as they first appear because the amount of vapor that penetrates incomplete closures or tears far exceeds the amount of vapor that penetrates intact fabric (Hanzelka et al., 1996).
- The MIST is a system level test of chemical protective ensembles (CPEs). The MIST and BRHA require a simulant only for generating protection factors for the CPEs.

- MeS has low toxicity.

MeS was chosen as the simulant for the MIST program. A single challenge level for the simulant was set at 100 mg/m^3. However, comparisons with other surrogates should also be developed to confirm that MeS is the most appropriate simulant.

MeS can provide a relative ranking of the vapor protection of various CPEs. However, comparative skin penetration rates for MeS and chemical agents, which would be required to relate MIST data to physiological endpoints, cannot be determined.

4

Test Methods and Sampler Selection

PURPOSES OF MONITORING

The MIST (man-in-simulant test) program is intended to establish protective factors for a personal protective equipment ensemble. Estimates of possible battlefield concentrations of chemical agent, the BRHA (body region hazard analysis), estimates of the cutaneous toxicity of agents, and personal factors, such as heat loading, may ultimately provide a rational basis for determining the relative acceptability of chemical protective ensembles. A protective factor is defined as the ratio of the concentration of the contaminant in air outside the suit to the concentration of the contaminant in air inside the suit. In the MIST program, the concentrations inside and outside the suit are measured at multiple sites by passive samplers. These concentrations can be determined by a variety of methods, which are discussed below. The precision and accuracy of any system is limited by its least precise or least accurate component. Measurements inside the suit will obviously be more difficult to take than measurements in the chamber and will be a more likely source of error.

The current monitoring system includes two layers of sophistication beyond determining a simple protective ratio, an estimate of the protection afforded to each region of the body and, ultimately, an estimate of the absorbed dose of simulant. Measuring protection by body region requires taking multiple samples at the same time from different places inside the suit. Although this creates a few special problems, it is still easier than estimating the absorbed dose of simulant.

Absorbed dose refers to the amount of simulant that penetrates the skin and enters the regional tissue or blood. The effective dose is the amount that actually reaches and injures living cells. Because the outer layer of skin consists of dead cells, simple skin exposure cannot be equated with poisoning. But there are many other routes to a living cell. For example, the toxicant may be ingested with food or water, it may be injected through the skin (carried on a shell fragment, injected by rupture of a pressure line, or deliberately injected), or it may be inhaled and absorbed through the large surface area of the lungs. Injection and ingestion are routes of exposure that are beyond the scope of the MIST program and are not considered here.

The absorbed dose differs from the effective dose because some of the agent that enters viable tissue or the blood stream may be destroyed before it reaches cells in the target organs. In addition, rates of detoxification may vary greatly among individuals. These differences can be genetically determined or modified by the environment (e.g., exposure to other drugs or chemicals).

Direct measurements of absorbed dose would require biological monitoring (i.e., drawing blood specimens, collecting urine, saliva, or breath samples, or taking tissue samples for analysis). Indirect estimates of absorbed dose require either simulating resistance to penetration by the skin or knowledge of how skin penetration varies with body region. The skin acts as a barrier in several ways. First, the keratinized epithelium (the layer of dry dead cells on the surface) acts as a passive barrier. The effectiveness of the passive barrier depends partly on the lipid solubility of the simulant but also on the thickness of the keratinized layer and the degree to which the skin is populated with sweat and sebaceous glands, which allow easier access to the blood. The thickness of the living layers of tissue below the keratinized layer also varies widely. Differences in thickness have a slight effect on penetration by simply increasing the opportunity for passive diffusion but may have a greater affect by increasing the hydrolysis and metabolism of the simulant during passage.

RATIONALE FOR THE SELECTION OF MONITORING METHODS

The monitoring methods for both chamber and in-suit concentrations should be precise and accurate. Ideally, monitoring methods should report concentrations to within a few percentage points of the true concentration of the simulant. In fact, the concentrations in the test chamber may vary more than a few percentage points; and the variations among suit trials and different wearers may greatly exceed the error introduced by monitoring. Nevertheless, precise and accurate measurements should be attainable within the design constraints of the test program and should be one of the most important criteria for choosing monitoring systems.

A monitoring method should not introduce errors. The chamber monitors should not alter the airflow within the chamber in such a way as to cause the concentrations to vary in different parts of the chamber. The in-suit monitor must not compromise the integrity of the CPE. Sampling systems that penetrate the suits alter the barrier properties of the suits or alter the form, fit, or function of the suit in other ways and could influence test results. Tubes that penetrate the suit to draw air from sampling points are the most obvious example of a monitoring system that alters the suit integrity. The barrier properties of a suit might be altered by adhesive tape used to affix samplers to the outer shell or by belts that compressed the sorbent layers or held them against the body allowing them to absorb perspiration. Alterations in the form, fit, or function most often apply to masks but could also affect suit performance. For example, bulky objects under a protective garment could increase the "bellows effect" whereby contaminated air enters the suit during some motions and leaves during others. Relatively small exchanges, if repeated often enough, can rapidly equilibrate concentrations inside and outside a garment. Bulky samplers under the garment could stretch neck, wrist, or ankle openings and degrade suit performance.

The weight and bulk of sampling equipment that a subject carries under the suit must not change activities or the level of effort required to accomplish tasks. Thus, the subject could not carry sampling pumps next to the skin, even though they would probably give the most precise measurements.

Many studies have shown dramatic differences in the absorption of chemicals through the skin of various body regions. In other words, absorption through the skin, unlike absorption through the lungs, is not simply proportional to the vapor concentration. Therefore, it may be desirable for the MIST program to simulate at least some skin penetration and absorption properties with the sampler rather than attempting to estimate the absorption from air concentrations.

CHAMBER MONITORING

The concentration in the chamber (outside the suit) is determined by the airflow rate into the chamber and the rate of simulant release into the air. According to the MIST test plan (Hanzelka et al., 1996), surfaces of the test chamber are stainless steel. Air flows through the chamber at 3.2 to 16.1 km/hr (2 to 10 mph or 0.89 to 4.5 m/s). The concentration is monitored by fixed samplers, called miniature infrared analyzers (MIRANS), located on the upwind end of the room and on the right and left sides of the inflow wall. Additional miniature automated chemical air monitoring systems are located in antechambers but are not used to monitor concentrations in the chamber. The detailed test plan does not indicate how the simulant is introduced into the air stream, what mechanisms are used to assure uniformity of simulant concentration within the air stream, or how much of the front wall is occupied by the inflow ducts. The MIST plan provides no information about the precision or accuracy of the MIRANS.

Supplying air with a fairly constant concentration of simulant over an extended period of time, in the absence of a mechanism for removing or absorbing significant amounts of the simulant, would tend to create a uniform concentration in all parts of the chamber. Maintaining uniform concentrations in animal exposure chambers, however, has proved to be a problem in toxicology experiments (Smith and Fowler, 1985).

The simulant is removed from the chamber air by respiration through the mask filters at a rate equal to the concentration in the air times the respiratory minute volume. The breathing rate will vary with activity. The respiratory rate for moderate activity in a workplace setting is assumed to be 10 m^3/8-hr shift, or 20.83 l/min. Exact chamber dimensions are not given in the test plan, but if one assumes a flow path of 5 m, there would be more than 600 air changes per hour even at the lowest flow rates. Therefore, removal is unlikely to appreciably diminish even the local concentration at any point in the chamber.

The simulant is also removed from the air by adsorption in or on the chamber surfaces and especially on protective garments. That rate will clearly vary with the efficacy of the suit and the type and intensity of activity and the position of the subject within the chamber. Army officials indicated that the chamber is allowed to reach equilibrium before tests are begun (Malabarba and Fidele, 1996). This should adequately control for the deposition of simulant on chamber surfaces but not necessarily on the suits. The large number of air changes per hour, even at low flow rates, should preclude an appreciable effect of simulant adsorption to chamber surfaces or protective garments on the simulant concentration. Actual chamber measurements are the only way to confirm this.

As an alternative to chamber monitoring at various locations, the MIST program has adopted the ingenious approach of placing passive monitors, identical to those used inside candidate ensembles, at identical positions on the outside of an impermeable garment. The impermeable suit is then worn by another subject who mimics the motions of the subject testing the candidate protective ensemble. This allows for comparisons of simulant concentrations at specific locations around the suit (e.g., elbow, knee, etc.)

OPTIONS FOR IN-SUIT MONITORING

This section outlines three major approaches that might be used to determine the amount of simulant inside a protective ensemble. Advantages and disadvantages of each sampling system are considered.

Active Sampling Systems

Overview

Active sampling systems, whether internal or external to the suit, draw air through a sorbent at a measured rate. The sorbent removes most (ideally all) of the chemical of concern. The chemical is then removed from the sorbent by solvent extraction or by heating the sorbent until the chemical is released. A variety of chemical analysis systems can then be used to measure the total amount of chemical desorbed. The total is then divided by the volume of air drawn through the sorbent to approximate the concentration. Because some chemical is usually retained on the sorbent and cannot be measured, chemists usually refine the estimate by calculating the fraction of chemical recovered from sampling tubes that have been spiked with a known amount of the analyte and dividing the quantity observed in samples by that fraction. This mechanism adjusts for proportional errors (the slope of the recovery curve). Other measurements can be used to adjust the zero-point (the interception of the recovery curve).

Active systems can introduce errors through uncertainties or variations in the flow rate. The pump is one potential source of variation. If tubing connects the pump to the sorbent tube or canister, bends or kinks in the tubing may also introduce error. The pump rate of battery-powered pumps may diminish as the battery discharges, but that problem has been well studied, and industrial hygienists usually calibrate the pumps at the beginning and end of each sampling period. For small changes, an average flow may be used. Nickel-cadmium batteries usually fail along a gradual, nearly linear curve, with rapid decay in voltage and current near the end of the cycle life. By avoiding the late phase, an average flow rate will give a satisfactory estimate provided that the concentration does not vary greatly during the test period. With some analytes, especially particulates or aerosols, a mechanical barrier may be created by clogging of the sorbent bed, which may increase resistance and decrease flow rates during the sampling period. It is not anticipated that this problem will arise with MeS as the primary simulant.

With either active or passive systems, the recovery of chemical decreases as the sorbent becomes saturated. For that reason, the size of the sorbent bed, the sampling rate, and the sampling time must be chosen so that the sorbent remains well below its saturation point. The problem is simpler with active systems, in which retained quantities may be large as long as the sorbent binds the analyte tightly enough to prevent breakthrough (passage of analyte completely through the system).

Other sources of error with either passive or active systems include influence of temperature, humidity, and the presence of other vapor constituents. These factors can be assessed by controlled trials of the system prior to a study. Data for the passive samplers show small, predictable variations with temperature and little variation with humidity effect. The effect of humidity on active samplers would probably be greater than on passive samplers with nonporous membranes.

External Pumps and Tubes

A system with pumps, connecting tubes, and samplers on the outside of a suit, with tubes passing through openings into the suit, has the advantage of permitting the use of bulkier equipment with better control over flow rates, sorbent bed volume, and operating temperature of the sorbent bed. Instruments that take direct readings might even be adaptable for intermittent monitoring. Measurements of airborne chemical concentration, the sorbent-to-analysis phase of the measurements, should not be affected, providing that the simulant does not adhere strongly to the tubing. Because of its low volatility, however, adherence to tubing is a severe problem with VX analysis systems, making direct collection of VX on a sorbent essentially impossible. With VX, the monitoring problem has been solved by installing a conversion pad, which substitutes a fluorine atom for the N,N-diisopropylamino-2-ethyl-thiol group of VX (see Figure 3-1). Similar solutions might be found for testing low volatility simulants.

The major disadvantages of external monitors are: (1) they compromise the integrity of the chemically protective ensemble being tested; (2) they limit motion if a fixed or bench-top system is connected to the subject; and (3) they add weight and bulk if the subject carries portable samplers.

The MIST study is designed to monitor in-suit concentrations at a large number of body sites. Tubes running either through the shell of the suit or parallel to the skin through neck, wrist, waist, or ankle openings could easily introduce gaps that would allow contaminated air to penetrate in a way that would not occur with an intact suit. Holes in the shell could be more readily sealed than gaps at openings, but this would require either multiple openings or passing tubes through the waistline or other sealing points, which creates the risk of abnormal simulant migration between body regions. Tubes running through the suit to external samplers would need to be collected into a bundle to prevent tangling during activities. To prevent distortion in the fit, the bundle would need to be attached to the suit with adhesives or to the soldier with a belt or harness. Adhesives can alter the barrier properties of the suit either by increasing or reducing permeability and can alter the fit by pulling on and distorting the suit, potentially aggravating the bellows effect that moves air in and out of the suit with motion. Straps may alter the results in a number of ways: they could decrease the normal flow of air between body regions by constricting the space between suit and skin; they could alter permeation qualities by compressing the suit material; they could decrease the area exposed directly to simulant concentrations and thereby decrease the in-suit concentration; and they could increase sweat absorption by holding the suit next to the skin, which could alter suit permeability. Finally, an active sampling system draws make-up air from outside the suit into the suit, which would necessarily increase the amount of simulant beneath the shell. This might be partially corrected by introducing an amount of clean air that precisely matches the amount of air that was removed. The process, however, would alter the motion of simulant within the suit leading to artificially high or, more likely, abnormally low readings.

Although many of these problems could be overcome with sufficient ingenuity and effort, collectively they indicate a significant potential for altering the form, fit, and function of the suit in ways that are difficult to predict. Using a single vacuum line or pump with a manifold system to multiple sampling points, for example, would solve some of the problems but would introduce uncertainty in the volume of air pulled through each sampler. Balancing a manifold system is difficult even with rigid tubing and would be even more difficult with flexible tubing. If a fixed or bench-top system were connected to the subject, the connecting tubes would serve as a tether, limiting the activities the subject could perform. Several pumps could be carried within the weight limits of a field pack, but then exercise protocols that do not require use of the field pack would be severely limited. The bulk of external portable samplers would be less problematic than the weight but might create problems if exercise protocols required crawling. The design of active samplers would make it difficult to simulate the barrier properties of skin. An absorbed dose could not be approximated with this system.

The problems described here do not necessarily eliminate sampling with pumps external to thc suit. That strategy might still be the best option because of the physical constraints of active samplers and the inaccuracy of passive samplers. The precision and accuracy of the analytical results of active samplers would probably not compensate for their disadvantages.

Pumps and Sampling Systems inside the Protective Garment

As an alternative to penetrating the shell of a protective ensemble with sampling tubes, pumps could be worn between the garment and the skin. An advantage of this scheme is that the pump would exhaust air into the space inside the garment, so no make-up air would have to be drawn from the outside. A disadvantage is that a new error would be introduced by placing a clearing mechanism within the suit. If the amount of simulant removed on the sampling medium were small in comparison to the flux through the suit, however, the error would be negligible. The precision and accuracy of internal pumps could be comparable to external samplers.

Although the structural integrity of the chemical protective ensemble being tested would be maintained by sampling pumps hung inside the garment, the garment's form, fit, and function would all be altered significantly. Form would be altered by pulling the garment away from the body over the pumps, which could open gaps at the waist or neck. It would also pull the garment closer to the skin on the opposite side, increasing sweat loading. The fit could be altered in many ways at various places, especially at friction points (waist, groin, underarm, neck, etc.) Protective garments that are designed to be "form fitting" would make the introduction of air sampling equipment particularly difficult. Function could be altered by increasing the heat loading within the garment as the result of the heat generated by the pump. If the number of pumps required does not fit practically under the suit, compromises in the number of sample points or the number of tests might be necessary.

In general, the design of active samplers does not lend them to simulation of skin barrier properties. An absorbed dose could not be approximated by a physical system. Sampling with pumps under the suit might still be the best option if the inaccuracies of passive samplers or other constraints preclude their use. Reducing the number of sampling points per test with active sampling would be troubling, but the requisite data could be gathered using more tests. The effect of heat loading could be calculated from information about the rate of heat generation by the pumps. If only a few pumps are used, the added stress might be negligible.

Passive Samplers

Passive dosimetry has become very popular for personal monitoring in recent years (Soule, 1991). The dosimeters or monitors use Brownian motion to control the sampling process, enabling lightweight, low-cost personal monitors that do not require a power source. They rely on a concentration gradient across a static or placid layer of air or other medium to induce a mass transfer. The following equation, based on Fick's law, gives the steady-state relationship for the rate of mass transfer:

$$W = D(A/L)\,(C_1 - C_0)$$

where W is the mass transfer rate, D is the diffusion coefficient, A is the frontal area of the static layer, L is the length or depth of the static layer, C_1 is the ambient concentration, and C_0 is the concentration at the collection surface.

If an effective collection surface is chosen, C_0 can be essentially zero, so the mass transfer or collection rate is proportional to the vapor concentration C_l. Note that the units of *D(A/L)* are volume per unit time, the same as for volumetric flow in a pump monitoring system. The rate of sampling of the contaminant is then the product of the *D(A/L)* term and the average ambient concentration.

The precision and accuracy of the overall system depend on the sampling process and the analytical steps. Precision and accuracy of the sampling process depend on the measured exposure time, velocity, and temperature, whereas the precision and accuracy of the analytical steps depend on the calibration standards, properties of the collection media, and the analytical method. The potential effects of velocity and temperature distinguish this type of monitoring device from the conventional dynamic or flow monitor.

When *L* (dependent on the resistance of the film barrier in this type of sampler) is large compared with the average boundary-layer thickness, sampling is barely affected by velocity. For temperatures between 10°C and 31° C, the variance should be no more than 1.8 percent. At higher or lower temperatures, corrections may be required.

An internal Army memo from 1994 discusses the effects of several factors on accumulated mass. The temperature effect is about 0.1 percent/degree F. In a cavity-type passive collector, the ratio of the sampler length to its diameter must be greater than three so as to minimize the effect of convection within the chamber. At low tangential face velocity, *V*, the uptake rate decreases because of external resistance to mass transfer in the boundary layer. Diffusion resistance through the boundary layer is proportional to $V^{0.5}$ for laminar flow and $V^{0.8}$ for turbulent flow. A commercial passive sampler (DuPont PRO-TEK™) exhibits a marked decrease in sampling rate below a laminar flow velocity of 1,000 cm/min (0.17 m/s). At a face velocity of 12 cm/min (0.2 m/s), the sampler collects 20 percent less vapor than the predicted amount, regardless of boundary-layer effects.

Theoretical models, laboratory evaluations, and field studies agree that diffusion samplers provide a reliable measure of mean vapor concentration if the fluctuation frequency is at least five 5 seconds. Specifically, steady-state mass uptake is approached closely if the period of the concentration change is less than 1.4 times the mean residence time of vapor in the diffusion zone t_R, where $t_R = L^2/2D$. The DuPont PRO-TEK™, which has a diffusion path length of 0.95 cm, will attain steady-state uptake if the fluctuation period is less than nine seconds, assuming that the diffusion coefficient for MeS at 80°F (27°C) in air is about 0.05 cm^2/s.

The key selection criteria for the MIST passive sampler include: (1) projection into the under-the-garment air space is sufficiently small so that it accurately samples the unaltered vapor stream in the vicinity of the sampling device; and (2) the absorption velocity W_v for MeS under MIST conditions is between 1 and 4 cm/min, the range observed for skin uptake of agents. Orifice-based samplers for which $W_v = D/L$ would require a diffusion path length of 0.75 to 3 cm to achieve the desired W_v range with MeS. Although this exceeds the dimensional requirement, interposing a barrier membrane in proximity to the adsorbent with a minimal diffusion path would solve the problem.

Natick Sampler

The Natick sampler is a type of passive sampler developed at the Natick RDEC specifically to detect MeS vapor for the MIST. The Natick system clearly protects the integrity of the protective ensemble. The samplers are no thicker than a common adhesive bandage and are less than 1 inch square (see Figure 4-1). Solubility of MeS in high-density polyethylene (HDPE) was reasonably linear over concentrations from 0 to 135 mg/m 3 with a coefficient of (1-mg MeS/gm HDPE)/ (125-mg MeS/m3) = 0.0082 m 3 /gm HDPE. The weight of a 1-mil HDPE cover on a Natick sampler is about 0.018 gm, so at a chamber concentration of 150 mg/m3, the quantity of adsorbed MeS would be:

(0.0082-m3/gm HDPE)(0.018-gm HDPE/sampler)(150 mg/m3)
= 0.022 mg per sampler

After treating the polyethylene films by heating to 95°C to 100°C for 16 hours, permeability to carbon tetrachloride increased substantially. At 18°C, permeability increased from about 55 gm/m2-day before treatment to about 80 gm/m2-day after treatment. At 40°C, the corresponding values were 120 and 160. Army scientists recommend pretreating the films in all cases to assure uniformity.

Water vapor often competes with other chemicals for binding sites on the sorbent, decreasing the sorbent's capacity to absorb the test chemical. Ambersorb and Tenax show less variation with humidity than carbon, but Army officials assert that the humidity effect has been insignificant in the Natick RDEC sampler. In contrast, humidity increased the skin absorption of organic compounds with partition coefficients in the range of 0 to 3 (Hawkins and Reifenrath, 1984). The partition coefficients of many chemical warfare agents fall within this range. These findings lead to the obvious conclusion that the permeability properties of skin and polyethylene are different, especially for a nonporous lipophilic membrane like polyethylene, which does not readily transmit water to the sorbent layer.

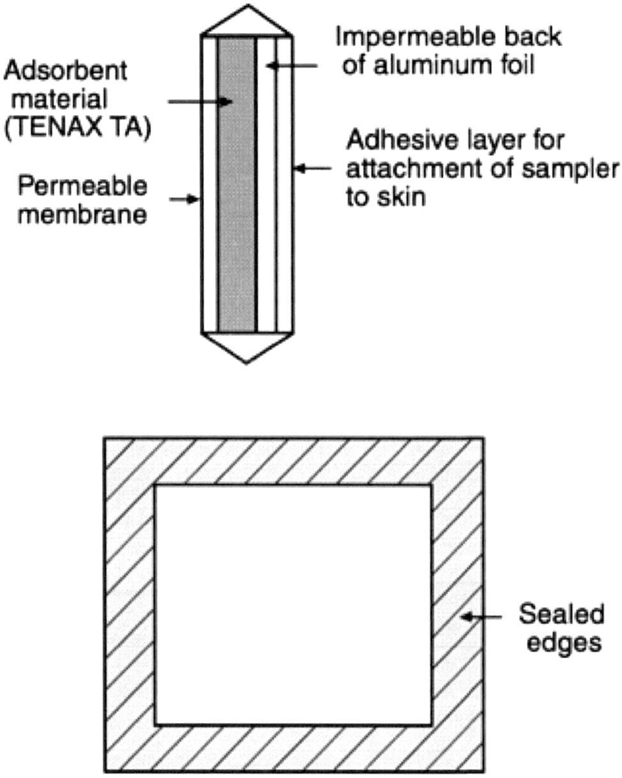

Figure 4-1 Diagram of Natick sampler.

The committee asked whether the same basic design would work with a different simulant. Army officials stated that the membrane has been tested successfully with a large number of chemicals. Confirming the effectiveness of a complete sampler would involve repeating some tests, but it seems likely that a suitable sorbent could be attached to the membrane and the design used to test other simulants.

5

Assumptions and Limitations

TEST PROTOCOL CONSIDERATIONS

An important aspect of the technical review of the MIST methodology is identifying all assumptions in the BRHA model. These assumptions must be closely examined to establish their impact on the design and implementation of the model. The resulting limitations must also be identified and evaluated to establish their impact on the applicability of the model to real-life situations. The committee's analysis of the test protocol raised the following questions.

The basic measurements of MIST/BRHA are protection factors for test suits for various anatomical sites against a 12,000 mg/m^3-min (concentration *x* time [*Ct*] factor) exposure to MeS (methyl salicylate). The protection factors are used to derive effective *Cts* against VX and HD. The MIST/BRHA can be used to rank the relative protective value of test suits against VX and HD but should not be used to predict physiological effects. For example, it would be wrong to suggest that wearing a given protective suit in a VX exposure of 3,225 mg/m^3-min (see Table 2-7) would result in symptoms of VX poisoning associated with a 70 percent depression in red blood cell cholinesterase. The data do not support a calculation of agent percutaneous absorption from the mass of MeS collected in passive samplers attached to the skin.

Given an effective *Ct* of 25 mg/m^3-min for VX exposure to produce a 70 percent depression in cholinesterase in an unprotected person, and 1,000 mg/m^3-min for HD to produce severe burns on an unprotected forearm, a Ct of 12,000 mg/m3-min in the MIST would be considered a massive challenge, especially for VX (if 1 mg of MeS is taken as the equivalent to 1 mg of chemical agent). Because all exposures were at a *Ct* of 12,000 mg/m^3-min, suit rankings might be different at lower *Ct* challenges.

The selection of MeS as a simulant for both VX and HD was based on its historical use as an agent simulant and on its safety for human use (see Chapter 3). One must ask, however, whether MeS is a reliable simulant of chemical agents. How does the diffusivity, solubility, surface tension, wetability, etc., of MeS compare with chemical agents? The physicochemical properties of VX, HD, and MeS are very different. MeS has a vapor pressure approximately 2 orders of magnitude higher than VX. Also, we do not know the relative distributions of these chemicals in system-level tests of chemical and biological protective suits.

The BRHA is based on the assumptions that the regional variation in VX skin toxicity is the same for liquid and vapor exposures and that the relative regional variation in skin toxicity to HD is the same as for VX. These assumptions have not been tested. For some compounds, some studies have shown that the regional variation in skin permeability appears to be compound dependent (Table 5-1); other studies have shown that permeability coefficients for liquid and vapor exposures are different (Barry et al., 1984).

Based on the statistical analyses of local and systemic effective *Cts* (Tables 2-5 and 2-7, respectively), approximately 75 percent of the

TABLE 5-1 Regional Variations in Human Skin Permeability as a Function of Test Substance

	Relative Permeability			
Anatomic Site	Hydrocortisone	Parathion	Malathion	Benzoic Acid
Forearm (ventral)	1.0	1.0	1.0	1.0
Forearm (dorsal)	1.1	-	-	-
Foot arch (plantar)	0.1	-	-	-
Ankle (lateral)	0.4	-	-	-
Palm	0.8	1.4	0.9	-
Back	1.7	-	-	0.8
Abdomen	-	2.2	1.4	1.6
Scalp	3.5	3.7	-	-
Axilla	3.6	7.4	4.2	-
Forehead	6.0	4.1	3.4	3.0
Jaw angle	13.0	3.9	10.0	-
Scrotum	42.0	12.0	-	-

test suits have overlapping CIs (confidence intervals) (at 95 percent significance level) with the battle dress overgarment. The CIs may be accurate reflections of the performance of the test suits; however, because the suits were not all tested the same number of times, conclusions cannot be drawn about the relative merits of the suits or about the discriminatory power of the MIST/BRHA. One must question then whether the tests have been replicated adequately to draw statistically reliable conclusions.

Because the MIST procedure is expensive, the natural tendency is to minimize the number of replications. It might be more appropriate to screen suits more rigorously prior to the MIST and to subject only the most promising candidates to the MIST/BRHA with more replications. Data from the MIST/BRHA have undergone ANOVA (analysis of variance). Additional tests (e.g., Dunnett, Neuman-Kuels test) should also be employed for multiple comparisons of protective ensembles. It may be of value to analyze data from individual anatomic sites as well as data from different test subjects.

The passive detectors are intended to measure skin deposition of MeS. However, no data have been established to compare MeS skin deposition directly with passive detector deposition.

The MIST is an accelerated test. In other words, higher concentrations of simulant are used for shorter periods of exposure (two hours) than are specified and required for suit protection (24 hours). Is this trade-off of time/concentration justified?

Have the temperature and relative humidity in the test chamber been appropriately chosen and controlled? One might suspect that the barrier and permeation properties of materials are temperature and humidity dependent; thus, performance must be evaluated under appropriate, relevant conditions. In the MIST procedure, is MeS introduced in the appropriate concentration (vapor challenge) to simulate a realistic chemical and biological threat? Are the convective dynamics (airflow rates) realistically reproduced and adequately controlled in the test chamber?

HUMAN FACTORS CONSIDERATIONS

The committee has two concerns about human factors associated with the test operations procedure for the MIST that should be addressed. The first deals with the closures of the protective garments and the second with the physical exercise routine to simulate field conditions under which the soldier would be expected to function.

The test operations procedure ensures that garments are properly closed at the beginning of each test. However, a complete, reliable interpretation of test results requires knowing the degree to which the closures remain closed during the test. The closures could be checked by the test supervisors at the end of the 120-minute exposure period when they check the positions of the passive samplers. Information about the closures would be helpful for interpreting differences in absorption levels at different anatomical sites. A related issue is the probability that soldiers would keep the garments closed under real combat conditions. The test procedure also specifies that an interview be conducted and a human factors questionnaire be completed by the test subject at the conclusion of the test. The interview and questionnaire could be critical to determining whether the suit would be worn as intended, with full closure, to evaluate test results from a practical standpoint.

The physical exercise routine in the MIST protocol may not adequately simulate field conditions, if for no other reason than that field conditions cover a wide range of variables and are almost impossible to specify. Nevertheless, the physical exercise must be rigorous and as reflective of anticipated field conditions as possible. The extent to which the movement disturbs the suit closures could be very important. Furthermore, body heat generated during exercise would increase the likelihood that a soldier might loosen the garment. The perspiration level during exercise is also important because perspiration changes the distribution of chemicals on the skin surface. It would be advantageous to review relevant literature comparing the types of exercises used in the MIST with actual field conditions to ensure that field conditions are adequately simulated.

ACETYLCHOLINESTERASE INHIBITION AS A BIOLOGICAL MARKER

Acetylcholine is neurotransmitter released at many autonomic nerve endings that binds to neurons and causes them to fire. Acetylcholinesterase is the enzyme that breaks the acetylcholine bond and returns the neuron to the resting state. Certain nerve toxins have long been believed to inhibit neural acetylcholinesterase enzyme activity based on the demonstrated ability of these agents to decrease acetylcholinesterase activity *in vivo* and the observation of the continued firing of neurons that could be explained by the action of these toxins. Understanding how known cholinesterase inhibitors work has proven to be very useful for developing neurotoxic pesticides and diagnostic tests for humans affected by them. These diagnostic tests are based on the presence in human red blood cell membranes of readily measurable levels of acetylcholinesterase activity, as well as the presence in plasma of a related enzyme known as pseudocholinesterase. However, in recent years a debate has developed about the usefulness of blood cholinesterase activity as a biological marker to predict neurotoxic effects.

Biological markers can be divided into markers that indicate the amount of exposure and markers that indicate presumed susceptibility. There is a continuum between markers of exposure and markers of susceptibility, and some markers can be classified as both. Interpreting biological markers depends on understanding the toxicology of the chemical, including its absorption, distribution, metabolism, excretion, and toxicity to the target organ.

Blood cholinesterase activity, either red blood cell or serum, should be considered both as an exposure marker and an effect marker. An example of an ideal marker of both exposure and effect is carboxyhemoglobin. Carbon monoxide bound to hemoglobin is both an integrated measure of carbon monoxide exposure in the past 8 to 12 hour period, and, through our understanding of the mechanism of CO toxicity, a predictor of adverse consequences.

Blood acetylcholinesterase activity is also, to some extent, a marker of both exposure and effect. The major limitation is the obvious fact that the target organ of concern is the brain, but the measurements are of enzyme activity in the blood. Subtle variations in enzyme structure between different tissues must be taken into account. Another limitation is that blood cholinesterase levels may vary due to genetic and disease factors. For most individuals, however, blood cholinesterase activity is a suitable biological marker of exposure. It is also a very useful biological marker of effect, as long as it is recognized that observable effects, such as fasciculations, do not begin until there is perhaps a 40 to 50 percent decrease in cholinesterase. Thus a 20 percent decrease in enzyme activity can be a useful marker of exposure but is not a definitive marker of effect.

The Army has used the data from a study by Sim (1962), which detailed the amount of liquid VX required to cause a 70 percent depression in red blood cell cholinesterase following application to different body regions as a quantitative indicator of regional sensitivity to HD or VX. Because cholinesterase measurement is quite variable, however, the data alone cannot be used to assign regional differences in agent sensitivity. The data may also have been compromised by prior, incidental, exposures, such as those found at Rocky Mountain Arsenal where low red blood cell cholinesterase levels could not be correlated with test exposure but were associated with carelessness in putting suits on and eating food placed on surfaces where used protective gloves had been placed.

BIOLOGICAL INTERPRETATION OF THE MIST/BRHA

The MIST generates an ensemble protection factor based on the ratio of simulant concentration outside the suit to the concentration inside the suit at each of many locations around the body (see Chapter 2). A protection factor for each specified location or region of the ensemble is calculated as the ratio of simulant detected in the absence and presence of the ensemble. Because component or swatch tests have been used to eliminate ensembles constructed from unacceptable materials, the MIST is particularly useful for detecting leaks around seams and closures.

The BRHA, combined with the MIST, simply weights the mass of simulant collected at a particular anatomic site by the surface area of a given skin region and the estimated regional variation in human skin permeability to chemical agent vapor. The results of the MIST/BRHA are still based on protection factors and require knowledge of the regional variations in skin penetration by the agent vapor. Currently, BRHA estimates of regional variations in VX and mustard vapor penetration are based on the data from Sim (1962), who studied droplets of liquid VX in humans.

The only way to validate the BRHA is through direct measurements of VX and mustard vapor penetration on excised human skin from different anatomic sites. An apparatus used at Dugway Proving Ground for generating stable vapor concentrations in swatch tests could be adapted for studies of excised skin. Regional variations in skin penetration, based on the small amount of data currently available, depends on the compound in question (Wester and Maibach, 1989) and may not even be relevant to vapor exposures (Barry et al., 1984) because most of the studies used solvents with the test chemicals. Sufficient data are not available to support the use of regional variations in transepidermal water loss in humans or regional variations in pesticide absorption in animals as measures of regional variations in skin absorption of agent vapor in humans.

Translating data from the MIST/BRHA into the biological effect of a simulated agent exposure (physiologic endpoint) will probably require developing a simulant for each chemical agent of concern because VX, mustard, and soman, for example, have different physical properties. Passive detectors may need to be modified or abandoned altogether because no artificial membrane has yet been shown to simulate the differential permeability of the skin and its response to changing temperatures and humidity. Noninvasive measurement of a simulant in the stratum corneum or the measurement of simulant and metabolite in urine or saliva may be more practical. The experience gained from monitoring civilians exposed to pesticides and other chemicals should be used to advantage. For example, scientists at the National Institute for Occupational Safety and Health have outlined a protocol for validating diffusive sampling techniques in the laboratory and in the field (Cassineili et al., 1987).

6

Conclusions and Recommendations

The MIST was developed to evaluate individual chemical protective suits. The MIST procedure is designed to compare the effectiveness of chemical protective garments and assess the operational requirements for protective garments. BRHA (body region hazard analysis), which complements the MIST, is an attempt to take into account regional body sensitivities to a chemical agent that has penetrated the protective garment. Together these models attempt to provide a quantitative measure of the effectiveness of a chemical protective ensemble under realistic dynamic conditions. The MIST/BRHA must be considered as procedures for evaluating the performance of complex systems, intended not only to characterize overall performance but also to identify the weakest elements in the system.

The committee understands that the MIST protocol includes the necessary procedures for data collection for evaluating the performance of candidate protective ensembles but *excludes* the site-specific analysis of data needed for the complementary BRHA. Thus, protection factors for protective ensembles can only be derived from the MIST.

However, the BRHA modifies the protection factor by introducing a consideration of the surface area and relative sensitivities of different body regions. The result of the MIST/BRHA remains a protection factor, but it cannot be used to derive a physiological interpretation of the data, which requires the translation of data for simulant disposition into an absorbed dose of agent.

In response to the charge to perform a technical assessment of the MIST program, the committee developed the following specific conclusions and recommendations. The committee also developed the general conclusions and recommendations presented at the end of this chapter.

SPECIFIC CONCLUSIONS AND RECOMMENDATIONS

TASK 1. Review the test methodology for the man-in simulant program.[1]

Conclusion 1. The MIST is a well-designed test protocol for evaluating chemical protective ensembles. However, the committee found that the test methodology was not based on preliminary testing that would eliminate ensembles with gross defects and allow more replications of tests be done on fewer candidate protective ensembles, thereby increasing the statistical power of the results.

Recommendation 1. The Army should screen ensembles prior to a full-blown MIST by video imaging the skin of test subjects after exposure to a fluorescent tracer or other physical tests. Screening should also include variations in ambient conditions (temperature, humidity, wind, and, perhaps, rain), activities (kneeling, sitting, and crawling), and sweat-soaked and dry test challenges.

TASK 2. Review the use of biological markers (e.g., cholinesterase inhibition) to predict the signs and symptoms associated with exposure to nerve (VX) and vesicant (HD) agents.

Conclusion 2. Body region hazard analysis (BRHA) is an innovative approach that takes into account regional variations in skin sensitivity to chemical agents. Although the basic approach is sound, the committee has the following reservations:

- A direct relationship has not been established between cholinesterase depression and the percutaneous absorption of agent.
- The relationship between liquid and vapor absorption has not been determined.
- BRHA was based on the local absorption of VX and may not accurately predict the absorption of HD.

[1] The original statement of task for Task 1 included "and the rationale for using methyl salicylate as a chemical agent simulant in this test program." The committee felt that this aspect of the review was reiterated in Task 4 and has addressed the question there.

- BRHA does not account for functional impairments from mustard-induced lesions in various body regions.
- BRHA does not account for individual differences in sensitivity to chemical agents.

A direct determinant of the toxicity of a chemical agent is the permeability of the skin by that agent at a given anatomic site. Therefore, the committee concluded that rather than basing the BRHA on highly variable indirect measures (cholinesterase depression) and assumptions, a protocol should be designed to quantify the *in vitro* agent permeability of excised human skin samples from different body regions. These techniques are well established and well accepted and could also be used to compare simulant uptake by human skin and passive samplers. Large differences may indicate a need to redesign the samplers. The vapor uptake of agent and simulant could also be determined for human skin and passive samplers. Large differences in the behavior of agent and simulant may warrant the selection of a different simulant or adjustments in the methods used to calculate protection factors.

Recommendation 2a. The Army should measure regional variations in skin penetration for HD, VX, and simulant vapors using excised human skin harvested from various anatomic sites.

Recommendation 2b. As a supplemental validation of the systematic BRHA, a biomonitoring protocol should be developed for the MIST, analogous to the protocol used to monitor pesticide exposures to agricultural workers. If the appropriate simulant is used, the calibrations obtained from *in vitro* studies could be used to relate suit performance to physiological effects based on the absorbed dose.

TASK 3. Review the test methodology for employing passive and active vapor and aerosol samplers during simulant tests and assess the data collection and analysis plan.

Conclusion 3. Passive samplers are appropriate for testing for the presence of vapor. The protocol, however, may not be valid for aerosols because the disposition of chemical agents in aerosol and vapor forms can be quite different. From the information recorded in the documents given to the committee for review, the committee could not confirm the uniformity of simulant concentration within the test chamber. Variations in concentration outside the protective ensemble could lead to errors in assessing the protective qualities of the suit.

Although passive samplers are generally regarded as less accurate than active samplers in bench trials, the differences in the results are small. The precision and accuracy of the Natick sampler is adequate for the intended purpose. The small size of the Natick sampler enables testing under the suit without incurring a number of disadvantages (outlined in Chapter 4) that would be incurred with active sampler pumps either inside or outside the suit.

A residual disadvantage of passive samplers may be a lack of sensitivity to brief variations in concentration, which would be of interest only for identifying the body positions or activities associated with leakage. Conventional active samplers would have the same disadvantage, but external samplers connected to a near-real-time monitor could provide this information.

Recommendation 3. Agent uniformity in all parts of the test chamber throughout the duration of the tests should be documented. In addition, concentrations inside the suit could be monitored with either active or passive samplers, despite logistical problems. Comparing simulant levels in the passive sampler with samples recovered from the stratum corneum of test subjects (the outermost layer of the skin, which can be removed by repeated applications of adhesive tape) would provide insights into sampler performance.

TASK 4. Determine whether the current chemical simulant methyl salicylate or an alternative simulant should be used in the MIST program.

Conclusion 4. Methyl salicylate is an appropriate simulant for the transport of chemical agent into protective ensembles. However, biological interpretations of the MIST/BRHA using methyl salicylate are not warranted.

Recommendation 4. Additional studies should be undertaken to establish absorption and transport properties of the simulant relative to the properties of the agents. *In vitro* studies using excised skin and mannequin studies (capable of simulating a bellows effect) can be used to accomplish this objective. With the appropriate consent, and oversight of a human use committee, excised human skin can be used for research. Samples can be obtained from cadavers or from surgical samples (e.g., abdominal skin, facial skin, etc.) Large differences in distributions may warrant using an alternative simulant.

GENERAL CONCLUSIONS AND RECOMMENDATIONS

General Conclusion 1. The first step in chemical and biological defense strategy is early detection and warning to provide situational awareness and permit steps to be taken to avoid the exposure of personnel and equipment. The complement to detection is protection. Chemical protective ensembles, as well as collective filtration systems and shelters, are used to insulate personnel from chemical and biological agents. Modeling chemical protective ensembles is a daunting task, and the Army's efforts to develop the MIST/BRHA should be commended. Modeling and simulation technologies are invaluable tools for training for operations in a chemical and biological warfare environment. They provide material and equipment design parameters and enable field commanders to integrate and interpret real-time data. However, deriving physiological endpoints from the MIST/BRHA is a complicated process that will require cooperation among the Army's scientists, as well as significant input from academia and industry.

General Recommendation 1. The development of new test methodologies should be done separately from routine ensemble testing. Once the criteria for suit performance have been established, decision points should be entered in a flow chart to reveal where additional work is needed. As of this writing, the Army has not adopted a clear approach to establishing physiologic endpoints from protective ensemble testing. However, this is an achievable goal that should be pursued to protect soldiers.

General Conclusion 2. The Army should ensure better cooperation among various disciplines (i.e., chemistry, toxicology, engineering, human factors, etc.). For example, scientists in CBDCOM's toxicology division have not participated in any significant way in the development of ensemble test methods.

General Recommendation 2. More integration between the various groups and technical disciplines will be essential for the development of future testing methodologies. All relevant parties should participate in the planning phase with the objective of reaching a consensus on research objectives, design procedures, analysis, and documentation.

References

Arca, V.J. 1996. Comparison of adsorbance for use with methyl salicilate. Report to the Commander. Dugway, Utah: Dugway Proving Ground.

Barry, B.W., S.M. Harrison, and P.H. Dugard. 1984. Correlation of thermodynamic activity and vapor diffusion through human skin for the model compound, benzyl alcohol. Journal of Pharmacy and Pharmacology 37: 84–90.

Cassinelli, M.E., R.D. Hull, J. Crable, and A. Teass. 1987. Performance of sulfur dioxide passive monitors. American Industrial Hygiene Association Journal 46(10): 599–608.

Cataldo, D., M.W. Aigotke, S.D. Harvey, R.J. Fellows, and S.W. Li. 1994. Acute Environmental Toxicity and Persistence of Methyl Salicylate: A Chemical Agent Simulant. PNL-9818-TD-2098. Aberdeen, Md.: Edgewood Research, Development and Engineering Center.

Conkle, J.P., R.E. Miranda, J.R. Fischer, R.W. Page, and D.L. Bartlett. 1986. Chemical Defense Collective Protection Technology: Vol. 1: Effects of Airlock Dimension, Clothing, and Exposure Concentration on Vapor Transport. USAFSAM-TP-86-2. San Antonio: U.S. Air Force.

Daugherty, M.L., A.P. Watson, and T. Vo-Dinh. 1992. Currently available permeability and breakthrough data characterizing chemical warfare agents and their simulants in civilian protective clothing materials. Journal of Hazardous Materials 30: 243–267.

Dugway Proving Ground. 1994. Agent/Simulant Relationship Study. Final Report WBS-01. Dugway, Utah: Dugway Proving Ground.

Environmental Protection Agency. 1992. Dermal Exposure Assessment: Principles and Applications. Interim Report EPA/600/8-91/001B. Washington, D.C.: Environmental Protection Agency.

Fedele, P. 1996. Presentation of MIST/BRHA to the Committee on Program and Technical Review of the U.S. Army Chemical and

Biological Defense Command. National Research Council, Aberdeen Proving Ground, December 1996.

Fedele, P.D., and D. Nelson. 1996. The Body Region Hazard Assessment Process. Paper presented at the Edgewood Research, Development and Engineering Center, Aberdeen, Maryland, December 12, 1996.

Fenske, R.A., S.W. Horstman, and R.K. Bentley. 1987. Assessment of dermal exposure to chlorophenols in timber mills. Journal of Applied Industrial Hygiene 24: 438–444.

Franklin, C.A., R.A. Fenske, R. Greenhalgh, L. Mathieu, M.V. Denley, J.T. Leffingwell, and R.C. Spear. 1981. Correlation of urinary pesticide metabolite excretion with estimated dermal contact in the course of occupational exposure to glutathione. Journal of Toxicology, Environment and Health 1: 715–731.

Hanzelka, J.A., T.D. Cronin, S. Dorman, S.L. Brimhall, and J.L. White. 1996. Detailed Test Plan for the Man in Simulant Test (MIST) of the Joint Services Lightweight Integrated Suit Technology (JSLIST) Program. Dugway, Utah: Dugway Proving Ground.

Hawkins, G.S., and W.G. Reifenrath. 1984. Development of an *in vitro* model for determining the fate of chemicals applied to the skin. Fundamental and Applied Toxicology 4: S133–S144.

Kauppinen, T., and L. Lindross. 1985. Chlorophenol exposure in sawmills. American Industrial Hygiene Association Journal 46: 34–38.

Malabarba, D., and P. Fedele. 1996. Presentation of MIST/BRHA to the Committee on Program and Technical Review of the U.S. Army Chemical and Biological Defense Command, Aberdeen Proving Ground, December 1996.

Merck Index. 12th ed., 1996. Whitehouse Station, N.J.: Merck Research Laboratories.

Newhouse, M., D. McEvoy, and D. Rosenthal. 1978. Percutaneous paraquet absorption: an association with cutaneous lesions and respiratory failure. Archives of Dermatology 114: 1516–1519.

Reid, R.C., J.A.M. Prausnitz, and B.E. Poling. 1987. The Properties of Gases and Liquids. 4th ed. New York: McGraw-Hill.

Reifenrath, W.G. 1980. Shower decontamionation efficacy. In vitro determination. Report 86. San Francisco: Letterman Army Institute of Research.

Reutter, S.A., and J.V. Wade. 1994. Review of Existing Toxicity Data and Human Estimates for Selected Chemical Agents and Recommended Human Toxicity Estimates Appropriate for Defending the

Soldier. Technical Report. Edgewood RDEC-SP-018. Aberdeen, Md.: Edgewood Research, Development and Engineering Center.

Sax, N.I., and R.J. Lewis. 1987. Hawley's Condensed Chemical Dictionary. 11th ed. New York: Van Nostrand Reinhold Company.

Scott, W.R., and B.W. Pointer. 1990. Evaluation of the Vapor Protection Capabilities of the Jacket/ Trouser Interface on the Regulation Ground-Crew Chemical Defense Ensemble. USAFSAM-TP-89-13. San Antonio: U.S. Air Force.

Sim, V.M. 1962. Variability of Different Intact Human Skin Sites to the Penetration of VX. Technical Report 3122. Army Chemical Center, Md: U.S. Army Chemical Research and Development Laboratories.

Smith, J.E., and W.K. Fowler. 1985. Analytical Methods Development. Vol. 1. Experimental Studies. Final Report. Aberdeen Proving Ground, Md.: U.S. Army Toxic and Hazardous Materials Agency.

Soule, R.D. 1991. General Principles in Patty's Industrial Hygiene and Toxicology. Vol. 1 (4th ed.) pp. 113–115, G.D. Clayton and F.E. Clayton, eds. New York: John Wiley & Sons.

van Hooidonk, C., B.I. Ceulen, J. Bock, and J. van Genderen. 1983. CW agents and the skin. Penetration and decontamination. Pp. 153–160 in Proceedings of the International Symposium of Protection Against Chemical Warfare Agents, Stockholm Sweden, 6–9 June 1983. Umea: Foersvarets Forskningsanstalt.

Wang, R.G.M., C.A. Franklin, R.C. Honeycutt, and J.C. Reinert, eds. 1989. Biological Monitoring for Pesticide Exposure. American Chemical Symposium Series 382. Washington, D.C.: American Chemical Society.

Wester, R.C., and H.I. Maibach. 1989. Regional variation in percutaneous absorption. In Percutaneous Absorption Mechanisms-Methodology-Drug Delivery, R.L. Bronaugh and H.I. Maibach, eds. New York: Decker.

World Health Organization. 1990. Public Health Impact of Pesticides used in Agriculture. Geneva, Switzerland: World Health Organization.

Zar, J.H. 1984. Biostatistical Analysis. 2nd ed. Englewood Cliffs, N.J.: Prentice-Hall.

APPENDICES

Appendix A

Body Region Hazard Analysis

INTRODUCTION

The purpose of this appendix is to summarize the BRHA (body region hazard analysis) model showing its rigorous basis and highlighting areas where additional articulation of the current model or model development might be useful. The discussion is intended to support not only the practical evaluation of protective suits but also the analysis of the efficacy of using MeS (methyl salicylate) as a simulant. This discussion relies heavily on the excellent presentations to the committee by Army personnel (Fedele, 1996; Fedele and Nelson, 1995).

CONCEPTS

The practical goal of the BRHA is to convert the information derived from a multidimensional experimental testing plan into a concise measure of the relative protection value provided by a given candidate protective ensemble. The basic idea is to convert measurements of exposures at 20 different body locations into a single measure that accounts or the variability of both chemical exposure and relative sensitivity at each body location.

The BRHA analysis provides two distinct modeling opportunities. The first, already in practice, is reducing the MIST data to a concise measure of the protective performance of the suit. The second is to examine the model derivation from first principles, thereby exposing key physical properties and approximations that enter into, and allow reduction of, the governing equations, in order to assess the efficacy of using MeS as a simulant. The model could thus define experiments that should be conducted.

The essential mathematical starting point for the BRHA is the probability distribution for response to an exposure level. It is evidently well accepted that this is the normal function of the log of the exposure. Equation 1 summarizes this information in terms of *s*, the normal equivalent deviate, *n,* the natural log of the population response geometrical standard deviation, *M,* the exposure, and M_{50}, the exposure value at which half the population shows a physical reaction.

$$s = n\ ln(M/M_{50}) \tag{1}$$

The BRHA applies this concept to each body region *j*. To do so, the terms of Equation 1 are rewritten as *Mj,* the exposure in region *j,* and M_{50j}, the specific exposure for each body region *j* that alone causes the mean response. Two additional concepts are A_j, the surface area of the body region *j,* and ε_j, the transport efficiency in region *j*. The parameter e_j accounts for the transport from skin deposits or absorbed agents to physiologically active sites or, in the parlance of Equation 1, $M = \Sigma\varepsilon_j M_j$.

Regional sensitivities have been measured, but an experimental determination of the transport efficiencies has not been made yet. The transport efficiency can be eliminated from the mathematical analysis by exploiting the behavior of the response probability distribution. Because at $s = 0$, half of the individuals will show a response, Equation 1 shows that this occurs when $M = M_{50}$. Therefore, $e_j = M_{50}/M_{50j}$ when local body region exposures are used in Equation 1. This also allows Equation 1 to be rewritten as Equation 2, which accounts for the 20 different body locations:

$$s = n\ ln(\Sigma\ M_j / M_{50\ j}) \tag{2}$$

The use of Equation 2 in the BRHA is facilitated by relating M_j to the vapor exposure. Far from saturation, exposure is the integral of the rate of delivery over time. This is shown in Equation 3, where C_o is the external vapor concentration, *t* is the time, v_j is the absorption velocity, and A_j is the surface area:

$$M_j = C_o\ t v_j A_j \tag{3}$$

Equation 3 allows Equation 2 to be rewritten as:

$$s = n\, ln(\Sigma C_o\, tv_jA_j/M_{50j}) \tag{4}$$

An exposure index C_oT can therefore be calculated as the point where half of the individuals in a population will respond $s = 0$, which provides Equation 5 for the median response exposure:

$$C_oT = 1/(\Sigma v_jA_j/M_{50j}) \tag{5}$$

The BRHA compares exposures under protected and unprotected (bare skin) circumstances. These exposures can be written as Equations 6 and 7, respectively:

$$M_{pj} = C_oTv_{pj}A_j \tag{6}$$

$$M_{bj} = C_oTv_{bj}A_j \tag{7}$$

Critical dosages taken from studies in the literature result in an effective amount of absorption M_{50j} when the exposure is to bare skin at various body regions. This is a useful standard and can be defined by substituting M_{50j} for M_{bj} and C_ET_{50j} for C_OT in Equation 7. This is shown in Equation 8:

$$M_{50j} = C_ET_{50j}v_{bj}A_j \tag{8}$$

Equation 8 and Equation 7 combined show that $M_{bj}/M_{50j} = C_oT/C_ET_{50j}$.

The BRHA seeks to account for M_{pj} by using the empirically determined relationships between M_{bj} and C_ET_{50j}. To do so, it is convenient to define the local body region protection factor, $P_{fj} = M_{bj}/M_{pj}$. This allows Equation (5) to be rewritten in terms of the experimentally determined protection factors

$P_{fj} = M_{bj}/M_{pj}$ and their effects as summarized by C_ET_{50j}. The result is shown as Equation 9, which provides the mean challenge level for individuals using protective systems relative to dosages that influence individuals through unprotected bare skin:

$$C_oT_{50} = 1/[\Sigma\, 1/(C_ET_{50j}\, P_{fj})] \tag{9}$$

In using Equation 9, the values of C_ET_{50j} are obtained from referenced toxicity studies involving human exposure to liquid VX (Sim, 1962). The application of these notions is slightly different for local and systemic analysis.

For local analyses, the P_f values for each location j are multiplied by the locationõs cutaneous mustard vapor toxicity value. The lowest vapor challenge level defines the effectiveness of the protective system. For systemic analyses, the critical whole body exposure $C_o\ T_{50}$ is calculated from an area-weighted variant of Equation 9 with $P_{fj} = 1$ for all j, as shown in Equation 10:

$$1/(C_O T_{50})_C = \Sigma\ A_j(\%)/(C_E T_{50j}) \tag{10}$$

The resulting $C_o T_{50c}$ has a value of 2.45 using Sim's (1962) data for A_j percent and $C_E T_{50j}$ represent the whole body exposure that, when uniformly applied to the 20 different locations on an unprotected person, results in the reaction.

The BRHA computes the actual challenge whole body exposure (or the whole body effective exposure discussed in the body of the report) by incorporating the location-specific, normalized exposures P_{fj} as in Equation 11:

$$1/C_O T_{50} = S\ A_j/(C_E T_{50j}\ P_{fj}) \tag{11}$$

The protective clothing is then rated with a protective actor PF as in Equation 12:

$$PF = C_O T_{50}/(\ C_O T_{50})_C \tag{12}$$

OPPORTUNITIES AND APPLICATIONS

BRHA calculations are executed in terms of a spreadsheet, which appears to be automated and systematic. Defending the logic of the BRHA model is more problematic. Perhaps the most basic question involves the normalization of exposures and the subsequent normalization of overall effects. Why is the starting point, although intuitively quite reasonable, rigorously correct? What is the physiological or transport basis for it? What are the essential approximations? Another question is the transformation of Equation 1 to Equation 2. It would be useful to articulate the rigorous logic on which these exposure "mixing rules" are based. Why normalize on a "spot-by-spot" basis before calculating the overall effect, for example? Why not use the actual exposures for each spot and then normalize? The weighting of the P_{fj} in Equation 11 is intuitively reasonable, but is it rigorously correct? Again, further articulation of the model derivation would be helpful.

Thus, the conceptual approximations, assumptions, and related limitations of the model would emerge from a more systematic derivation, complete with sample calculations. But another, more global benefit would also result. Systematic development would reveal the dependence of the model parameters on the physical properties (e.g., diffusivities, adsorption and absorption constants, viscosity, vapor pressures) of the system. This would, in turn, provide a more quantitative basis for assessing the efficacy of MeS as a chemical agent simulant.

NOMENCLATURE

A_j	surface area of body region *j*
b	bare skin
C_o	external vapor concentration
$C_E\mathrm{T}_{50}$	critical dosage that results in an effective amount of absorption (M_{50j})
C_0T	exposure index
$(C_0T_{50})_c$	critical whole body exposure for systemic analysis
M	exposure
M_j	exposure in region *j*
M_{50j}	specific exposure for each body region *j* that alone causes the mean response
M_{50}	the exposure value at which half the population shows a physical reaction
n	the natural log of the population response geometric mean standard deviation
p	protected skin
P_{fj}	the local body region protection factor
PF	rating factor for protective clothing
s	the normal equivalent deviate
t	time
v_j	absorption velocity

REFERENCES

Fedele, P. 1996. Presentation of MIST/BRHA to the Committee on Program and Technical Review of the U.S. Army Chemical and Biological Defense Command. Aberdeen Proving Ground, Md., December 1996.

Fedele P.D., and D. Nelson. 1995. A report on the Method of Assessing Full Individual Protective System Performance Against Cutaneous Effects of Aerosol and Vapor Exposures. Aberdeen, Md.: Edgewood Research, Development and Engineering Center.

Sim, V.M. 1962. Variability of Different Intact Human Skin Sites to the Penetration of VX. Technical Report 3122. Army Chemical Center, Md: U.S. Army Chemical Research and Development Laboratories.

Appendix B

Panel and Committee Meetings

PANEL MEETINGS

October 22, 1996 Aberdeen Proving Ground, Maryland

Objectives: Review statement of task; approve the project plan; approve the outline and study concept as articulated in the draft report concept; begin gathering data by discussing the man-in-simulant test (MIST) program with personnel from the Edgewood Research, Development and Engineering Center (RDEC) and the U.S. Army Center for Health Promotion and Preventive Medicine (USACHPPM); schedule writing assignments for chapters of the report.

Presenters

Joseph Vervier, Technical Director, U.S. Army Edgewood RDEC

James Baker, Chief Scientist, Office of the Technical Director, U.S. Army Edgewood RDEC

Janet Jensen, U.S. Army Edgewood RDEC

Sandra Thomson, U.S. Army Edgewood RDEC

John Ferriter, U.S. Army Edgewood RDEC

Ronald Crosier, U.S. Army Edgewood RDEC

Paul Fedele, U.S. Army Edgewood RDEC

Sharon Reutter, U.S. Army Edgewood RDEC

Douglas Nelson, USACHPPM

Steven Kistner, USACHPPM

March 12, 1997 Ft. Mitchell, Kentucky

Objectives: Provide panel members an overview of the role of the MIST; provide panel members perspectives of the organization involved in the MIST program; review and refine first full message draft; discuss report schedule.

Presenters

James Hanzelka, Protection Group Leader, Dugway Proving Ground
Donald Riven, Principal Scientist, Natick RDEC
Teresa Kocher, Test Integrator, Aberdeen Proving Ground
Douglas Bryce, Product Manager, Quantico, Virginia
Charles Gidley, Deputy Project Manager, Fort Belvoir, Virginia

FULL COMMITTEE MEETINGS

December 11–13, 1996 Aberdeen Proving Ground, Maryland

Objectives: Introduce and orient the new committee members to the National Research Council and the Edgewood RDEC; discuss the committee's goals, objectives, milestones, and statement of task for the new contract; discuss the background and approach for the planned assessments.

Presenters

Joseph Vervier, Technical Director, U.S. Army Edgewood RDEC
Harry Salem, Chief Scientist for Life Sciences, U.S. Army Edgewood RDEC
Paul Fedele, Team Leader/Simulation, U.S. Army Edgewood RDEC
D.G. Parekh, Team Leader/CB Detection, U.S. Army Edgewood RDEC
Randall Wentsel, Team Leader/Environmental Technology, U.S. Army Edgewood RDEC
Jeffrey Smart, Command Historian, CBDCOM

April 24–25, 1997 Washington, D.C.

Objectives: Provide standing committee members with an overview of current activities; update milestone chart for both panels; conduct composition and balance discussion.

MIST Panel

- Breakout session to write, review, and approve first full message draft

Presenters

Joseph Vervier, Technical Director, U.S. Army Edgewood RDEC

James Baker, Chief Scientist, Office of the Technical Director, U.S. Army Edgewood RDEC